ARTIFICIAL INTELLIGENCE

What will the future be?

A dystopian landscape controlled by machines or a brave new world full of possibilities?

Perhaps the answer lies with Artificial Intelligence (AI)—a phenomenon much beyond technology that has, continues to, and will shape lives in ways we do not understand yet.

This book traces the evolution of AI in contemporary history. It analyses how AI is primarily being driven by "capital" as the only "factor of production" and its consequences for the global political economy. It further explores the dystopian prospect of mass unemployment by AI and takes up the ethical aspects of AI and its possible use in undermining natural and fundamental rights.

A tract for the times, this volume will be a major intervention in an area that is heavily debated but rarely understood. It will be essential reading for researchers and students of digital humanities, politics, economics, science and technology studies, physics, and computer science. It will also be key reading for policy makers, cyber experts and bureaucrats.

Saswat Sarangi is a theoretical physicist by training with a PhD from Cornell University, Ithaca, NY, USA. After his PhD, he was a research scientist at Columbia University, New York City, USA. Saswat started his finance career in New York City working as a Quant at Bloomberg and later at Citigroup. He currently works in Cambridge, Massachusetts, USA.

Pankaj Sharma is an engineer from IIT Kharagpur, India, with an MBA from the Faculty of Management Studies, University of Delhi, India. He has more than 15 years of diverse work experience in various leadership roles with global investment banks, Indian equity brokerages, state-owned enterprises and start-ups. Pankaj turned full-time researcher in late 2016 to do in-depth work on contemporary issues. Earlier, he was a ranked Equity Analyst with UBS, Citi and JP Morgan. Pankaj has also been a regular contributor to print and electronic media. Pankaj published his first two books in 2017: *Demonetization: Modi's Political Masterstroke* and *2019: Will Modi Win?*

ARTIFICIAL INTELLIGENCE

Evolution, Ethics and Public Policy

Saswat Sarangi and Pankaj Sharma

LONDON AND NEW YORK

First published 2019
by Routledge
2 Park Square, Milton Park, Abingdon, Oxon OX14 4RN

and by Routledge
711 Third Avenue, New York, NY 10017

Routledge is an imprint of the Taylor & Francis Group, an informa business

British Library Cataloguing in Publication Data
A catalogue record for this book is available from the British Library

Library of Congress Cataloging in Publication Data
A catalog record has been requested for this book

ISBN: 978-0-8153-9682-6 (hbk)
ISBN: 978-1-138-62538-9 (pbk)
ISBN: 978-0-429-46100-2 (ebk)

Typeset in Bembo
by Taylor & Francis Books

For our Alma mater, Indian Institute of Technology Kharagpur

CONTENTS

ILLUSTRATIONS

Figures

Tables

PREFACE

The story behind this book

> I'd like to share a revelation that I've had during my time here. It came to me when I tried to classify your species and I realized that you're not actually mammals. Every mammal on this planet instinctively develops a natural equilibrium with the surrounding environment but you humans do not. You move to an area and you multiply and multiply until every natural resource is consumed and the only way you can survive is to spread to another area. There is another organism on this planet that follows the same pattern. Do you know what it is? A virus. Human beings are a disease, a cancer of this planet. You're a plague and we are the cure.
>
> —*Agent Smith in* The Matrix *(1999)*[1]

Mumbai is one of those mega cities where life never slows its pace. It was way past midnight when, sitting in a nondescript Mumbai cafe in early 2017, Saswat and I started discussing the recent turn of global events. A number of unprecedented events seemed to be steering the world towards an unpredictable future. And in our discussions one particular topic stood out as something with a significant ongoing and potential future impact on humanity, but barely getting sufficient coverage in media: Artificial Intelligence (AI). No matter how one looks at it, no matter how one slices and dices the topic and the scenarios, AI has a huge potential to alter the trajectory of human society. However, the extent of the lack of knowledge and appreciation among the general public about a topic with such potential is worrisome.

The first time I heard of AI, I thought it was something related to pretense. As in someone who is not a great musician but trying to pretend to be one, or someone who is not a very nice person but feigning to be one. It was as if I was equating artificial intelligence with fake intelligence. Eventually I found out what AI meant. Over time, especially over the last few years, I really started appreciating the extent to which AI can, and will, impact human lives. In the current day and age, one does not even need to try to grasp the extent. AI, in its various products and forms, presents itself in our daily lives in a way that is hard to ignore. A machine that has a good level of *intelligence* in one or more areas is clearly useful for humans. These machines are useful in offices, factories and in homes. And because of their ever-increasing prowess and expertise; these machines will become more commonplace in the future.

All sorts of possibilities exist for potential future applications of AI. Robots may become capable of measuring a human's emotional reaction to events and then might also be able to act on the assessment. It may try to align the displayed emotional reaction with the natural and appropriate emotional reaction by taking measures to assist and help the human subject. Perhaps, when a robot witnesses a human in pain, it will be able to administer medicine. Or, perhaps we will see robots offering psychological help to humans after measuring a human's emotional issues. AI possibilities are many.

All sorts of possibilities also arise for humans in a future with an ever-increasing presence of AI. Some consequences are already being discussed. Machines will certainly take over an increasing number of jobs that belong to humans. However, many aspects of machines' impact on humans is really an unknown. How will the new machine age impact humans psychologically? As we remain increasingly glued to smartphone screens, as virtual reality products present novel possibilities, will humans become a bit like the machines emotionally? In its attempt to make machines more human-like, will humans become more machine-like? How will the new machine age impact the fabric of society, the structure of family? Will there be a need to revisit the basic principles of ethics and morality as AI becomes commonplace?

But, some of these are larger philosophical questions. Apart from humans becoming less like humans, there are more tangible, practical and both immediate and long-term issues with AI that concern humankind. Will the current wave of heightened interest in AI sustain, or after the euphoria will we again be headed towards an "AI winter" just like in previous instances? Does AI have the potential to make mass unemployment a reality? Can AI become a threat for humankind as it becomes uncontrollable?

Perhaps it can become a huge disaster if it falls into wrong hands? Are governments around the world thinking enough about AI? How important is it to develop a comprehensive policy response and is it even practically possible to develop one? Why has AI not yet become a topic of discussion in mainstream media and has been limited to intelligentsia only? Why is the education and involvement of the general public in this discussion so important?

This book is an attempt to think clearly about some of the issues and hopefully to motivate the reader to think, talk, and find out more about AI and its implications. The idea is to make an effort to make the discussion on AI a bit more mainstream, and to take it from the confines of tech communities to the general public because there is hardly going to be anyone who will remain untouched and unaffected by AI.

—*Pankaj Sharma*

Note

1 *The Matrix* is a 1999 science fiction action film written and directed by the Wachowskis (credited as the Wachowski Brothers) and starring Keanu Reeves, Laurence Fishburne, Carrie-Anne Moss, Hugo Weaving and Joe Pantoliano. It depicts a dystopian future in which reality as perceived by most humans is actually a simulated reality called "the Matrix", created by sentient machines to subdue the human population, while their bodies' heat and electrical activity are used as an energy source. Computer programmer Neo learns this truth and is drawn into a rebellion against the machines, which involves other people who have been freed from the "dream world". *The Matrix* was first released in the United States on March 31, 1999, and grossed over $460 million worldwide. The success of the film led to the release of two feature film sequels, both written and directed by the Wachowskis: *The Matrix Reloaded* and *The Matrix Revolutions*. *The Matrix* franchise was further expanded through the production of comic books, video games and animated short films, in which the Wachowskis were heavily involved, and even inspired books and theories on ideas in religion and philosophy; www.imdb.com/title/tt0133093/quotes (Accessed on 22 February 2018), www.imdb.com/title/tt0133093/?ref_=ttqt_qt_tt (Accessed on 22 February 2018), www.imdb.com/name/nm0915989/?ref_=tt_trv_qu (Accessed on 22 February 2018).

ACKNOWLEDGEMENTS

We dedicate this book to the **Indian Institute of Technology Kharagpur (IIT KGP)**. The years spent by both of us at IIT KGP were not only our formative years, but also provided us with a magnificent world view and that too at such a young age when our minds were certainly more impressionable. There are so many good things about IITs, like a very competitive selection process, a world-class infrastructure, academic rigor and brilliant faculty, that it is impossible to select any one aspect. But still, if we had to choose just one thing, it would be the pool of students.

And this is not in the slightest bit because those 17 or 18 year olds are all good at academic work or have incisively analytical minds but more due to the fact that almost all of them are passionately interested in a field or two. Such a diverse and deeply passionate set of teenagers make the IIT platform a powerhouse of ideas and intellect, and that experience was so enriching that we never got that again. Thanks to all the teachers, staff, friends, department mates, hall mates, batch mates, seniors and juniors. Really heartfelt and sincere thanks to all of you!

We are immensely grateful to all the teachers, colleagues, professional associates, friends and family members who have not only influenced our thought process knowingly or unknowingly, but also helped in molding who we are, as people and as professionals. There are so many of them that it would be virtually impossible to remember each and every one but still, none of them would have a contribution that is not significant. We have a deep sense of gratitude towards each and every one of them.

Saswat would like to thank his wife, Sarmistha, and his son, Ishaan. Without their patience this project would not have been possible. Saswat would also like to thank his father, Damodar Sarangi, and his mother, Annapurna Sarangi, who always encourage him in his pursuit of knowledge. Saswat benefited immensely from the intellectually stimulating conversations with various experts and non-experts in Cambridge, Massachusetts, where one can find both AI experts and AI alarmists with equal ease. He would particularly like to thank Paul O'Connell for bringing to his attention work by Nick Bostrom. Saswat would also like to thank various contacts in the financial industry, various friends and acquaintances working in the sales and trading desks of Wall Street banks, with whom he had numerous discussions about the impact of automation. A number of these contacts had to, unfortunately, give way to automation and AI, as their jobs required less of the human touch and more of the AI finesse. Roy Rodenstein, with his deep knowledge of technology and his concerns about the human impact of AI, was especially helpful in sharing articles and ideas relevant to a number of topics discussed in this book.

Finally, the consent to go ahead and cooperation from immediate family is never a sufficient condition for an endeavor like this but is almost an absolute necessity. Pankaj is deeply thankful to his wife, Shikha and son, Pulin, for their patience and accommodation keeping in view the enormous demand on Pankaj's time this book had warranted during the background research and writing phase. It would not have been possible without the support from them. Pankaj is keen to acknowledge the contribution and encouragement received from so many people that the list below is really long but, still, incomplete. The first inspiration that comes to his mind was from his maternal grandfather, the late Shri Janardan Shastri "Avaneendra" to whom he would attribute his love for reading and writing. His friends (Sandip Bansal, Saurav Sanyal, Sunil Teluja, Niraj Khare, Rohan Padhi, Rupal Mehrotra, Rohan Arora, Pavitra Kumar, Ramesh Mantri, Sanjeev Soni and Unmesh Sharma), his teachers at Morar, Kharagpur and Delhi (the late Shri Kulwant Singh Sachdeva, Mahavir Sir, Sharma Madam, Bhupendra Sir, Rakesh Sir, Maheshwari Sir, Dr. Gokarn, Dr. Raheja, Dr. Sha, Dr. Ghosh, Dr.Satsangi, Dr.Narag, Dr.Mitra, Dr. Singla, Dr. Pandit and Dr. Kaur to name a few of them), his workplace superiors at different points of time (Suhas Hari, Suresh Mahadevan, Venkatesh Balasubramaniam, Srikant Bharti, Ashish Gupta, Ajay Garg, Gordon Gray and Kishore Gandhi), and his brothers who provided solid support in this journey (Ranjan Sharma, Neeraj Sharma and Pushkar Shukla). So too did his parents (Mr. Kamlesh Kumar Sharma/Mrs. Veena Sharma and Dr. Ram Gopal Shukla/Mrs. Mithlesh Shukla).

We wish to acknowledge our sincere gratitude to Mr. Aakash Chakrabarty, Ms. Brinda Sen, Ms. Sophie Watson, Ms. Katie Finnegan and other team members from Taylor & Francis, whose continuous support and encouragement made this book possible. We were never very good in making a realistic and fair assessment of what were the key areas of development for this book since the time we submitted the first version of our manuscript, but it was Brinda and Aakash who helped us at each step. The way both of them painstakingly reviewed each and every word, we have no hesitation in admitting that a thorough review of a text is almost as difficult as writing one and it adds immense value. We are also grateful to all other team members at Taylor & Francis for their unwavering patience all the time during which this book took shape.

INTRODUCTION

> By far the greatest danger of Artificial Intelligence is that people conclude too early that they understand it.
>
> —*Eliezer Yudkowsky*[1]

Warren Buffett is one of the most successful stock market investors in the world.[2] There might be an expert or two who would argue that Buffet is not the *GOAT* (Greatest of All Time) and that there are others more deserving of the title. But, even these critics would agree that no other stock market investors on Earth can match Buffet in the legends associated with him. These legends are related with his investing style, his life principles and his philanthropy. Even his affinity for drinking plenty of aerated drinks or his almost dogmatic aversion to modern gadgets, or his bridge sessions with Bill Gates make regular media headlines.[3]

One of these legends is his firm Berkshire Hathaway's Annual General Meeting (AGM), which for many of his devotees is an annual pilgrimage event. Once a year, Warren Buffett gives the shareholders of Berkshire Hathaway a chance to come and visit his home town, Omaha, to rub shoulders with him. People come from all over the world to watch Warren Buffett, and his deputy, Charlie Munger, chew candy, drink Coke, and answer questions from a few select shareholders. The fans of Buffett and Munger eagerly devour every pearl of wisdom that the duo shares at the event. Most often, and this is applicable to many of them, these fans come only to soak in the atmosphere, and hence they rarely go home disappointed.

About 20 years ago, a mere 10-year-old shareholder, Thomas Kamei, took the microphone in one of the Berkshire Hathaway's AGMs and asked Mr. Buffett about his thoughts on the internet and its potential impact on companies. Buffett said he saw a threat in the internet, but couldn't tell how it would impact his investments. *Fast forward to 2017...* Thomas Kamei, now a young man in his late twenties, was back at the microphone at this year's Berkshire Hathaway annual shindig. This time Kamei asked a similar question, but this time, the question was about AI. What does Warren Buffett think about the potential impact of AI? One wonders if Buffett remembers Kamei as the kid who asked him that question about two decades ago. ***But Buffet seemed to have more conviction in his answer this time. "AI is here to disrupt", said the Oracle of Omaha.***[4]

But, why should we listen to Buffett? He may be an expert in investments but that does not make him one in AI too. That is a fair point but Buffet is certainly not alone. Based on the recent evolution of AI research and development, this book takes the view that AI is indeed here to make major disruptive changes in our society. It ponders about what needs to be done by society and policy makers to prepare for a future with AI. The book looks at some of AI's history, at its spectacular progress, and a handful of AI's promises. More importantly, this book unapologetically expresses worry at our lack of preparation for a future rife with intelligent machines.

And whether we like it or not, this lack of preparation is not just limited to the uninformed and unaware among us. Even the Treasury Secretary of the United States, Steven Mnuchin, does not think that AI will take over US jobs for at least another 50 to 100 years. Such indifference among policy makers is a reason to worry. The history of the human race on this planet tells us that tribes, societies, or nations that survive and thrive do so not simply because they are stronger, but because they are better at adapting to changes. This has been a very fundamental principle and we should ignore it at our own peril. This applies equally to AI, its disruptive ability and our preparedness for a world full of AI-related technologies.

At the outset, we would like to add a disclaimer that this is not a technical book; this is not a book for computer scientists or researchers in AI. So, if you are looking for the latest in AI research, you are in the wrong place. This book is for those who might not have a technical background but who take the attitude "one doesn't need to be an auto mechanic in order to drive the car well". Someone who wants to understand – what AI is and what it is not; what AI can do; how AI might affect humans, society and the economy; can AI be an existential threat to humans; how can humans prepare for a world where AI will pay a huge role in their lives; is there anything we can do right now or is it already too late; or is this problem just too complicated leaving us as mere spectators to the arriving tsunami of AI?

The impact of technology on job creation in the economy is a hot topic, and almost daily we see something or the other in the media on this subject. Technology and AI are making many of the blue-collar and white-collar jobs obsolete in an irreversible way; this is simply an inescapable development. AI impacts not only jobs that are monotonous and which can be defined and programmed precisely, but also the ones which require thinking and judgment. Thinking and judgment are qualities that have been associated only with humans throughout humankind's history, but this status of humans will be sorely tested over the next few years with new AI research and faster-than-ever developments.

There is another dimension, and that is linked to economics. Why has AI suddenly become so much talked about? In many ways, AI and automation are leading to a scenario where "capital" is becoming the one and only "factor of production" that matters. "Labor" does not matter anymore in the overall scheme of things. And, this means more inequality and a huge social shift that is going to be unprecedented not just in decades but in centuries. This has serious implications for the world economy and human society.

Furthermore, the impact of AI is not only constrained to labor, productivity and employment. There are a number of other issues that revolve around the interaction of AI and human society, and we will discuss these issues over the course of the book.

- Is AI just like other technologies which swept through society in the past or is this time really different?
- Similar to nuclear energy and genetics, does AI also need an informed discussion on the ethics and "Dos and Don'ts" of it?
- Is there a need for a global and comprehensive policy to deal with AI? Or will a "hands-off" approach work better?
- How has the availability of cheap capital influenced investments in newer technologies, and has that been an influence in the rapid progress of AI?
- Why is it important for policy makers to start an informed and balanced discussion on AI technologies?
- Is there a point in time after which it becomes too late to take proactive measures to deal with AI?

Is the current popularity of AI only hype without substance?

Let us consider the opposite possibility. Is AI just the latest buzzword, more of a fad that will run out its course eventually? Is there a possibility that AI will once again retreat to the confines of university laboratories after promising goods but failing to deliver? Admittedly, a lot of speculation about the future of AI touches on very futuristic possibilities. Any prophecies about the future of AI are fraught with fantasy and uncertainty. A waxing and waning of interest and effort is natural for any scientific and technological endeavor. And, most likely, once AI research has solved the more tractable problems it currently faces, it will once again go through a phase where only the most hardcore enthusiasts persevere to solve the hardest problems.

However, it is undeniable that the social impact of the latest developments in AI has already been huge. Jobs have been ceded to AI which until now were considered impossible for a machine to perform. Machines are now available, either commercially or in laboratories, to intelligently clean your house, drive your car, read a book to you, write the newspaper articles which you will find interesting, diagnose your medical records when you fall sick and suggest the best line of treatment, or even to predict your future behavior. In fact, the possibilities are endless and opportunities infinite on what AI will be able to deliver.

There are fundamental reasons as to why more technology in the form of automation and AI is an attractive option for industries not just in the developed world, but also in emerging countries. Over the last couple of decades, outsourcing jobs to cheaper developing countries became a natural recourse for the developed world as local labor became more expensive. However, outsourcing came with its own logistics-linked and other social, personnel and practical challenges such as keeping track of overseas operations, working across multiple time zones and managing multiple teams over multiple locations. Over time, as wages increased in these developing countries, the economics of outsourcing also faced challenge. For a developed nation, automation and AI provide the possibility of bringing operations back home, avoiding the hassles of outsourcing.

Even for countries like China and India, large-scale automation provides an alternative to labor-related issues. China, which is the factory of the World, has invested heavily in introducing AI to manufacturing. Foxconn, one of the largest private employers in China, has been massively troubled by labor unrest issues over recent years; to resolve these challenges, it now has major plans to bring AI to the factory floor as a way to deal with some of these problems. In India, a number of information technology companies are increasingly using automation for routine tasks. This will translate to a drop in hiring freshly minted engineers, of whom more than 1.5 million are produced every year in India as per AICTE data.[5] We cannot be entirely certain of the full extent to which AI will impact jobs, and whether our response, in terms of development of a legal and policy framework, will be quick enough. But, AI is certainly here to stay and to grow its reach.

Over the long term, there is also the existential question that is increasingly being taken seriously. A number of experts believe that AI might eventually become superior to the human brain, and though we don't know when such a time will arrive, the existence of such a possibility itself is fascinating and intimidating at the same time. But, the impact on jobs from AI is a more immediate challenge and as far as this is concerned, the current simplistic versions of AI are already significantly affecting jobs. If some day AI does become sufficiently flexible and general to emulate human-like thinking, the impact on society will be immense.

What next?

AI research, like any other field that promises big, has seen cycles of ebb and flow in enthusiasm for what it can deliver. But, due to technological breakthroughs and commercial products that are influencing human lives in very tangible ways, once again hopes are high. Driverless cars, Siri and Cortana, IBM Watson, Deep Learning, Automated Trading, have all become household words. What has also

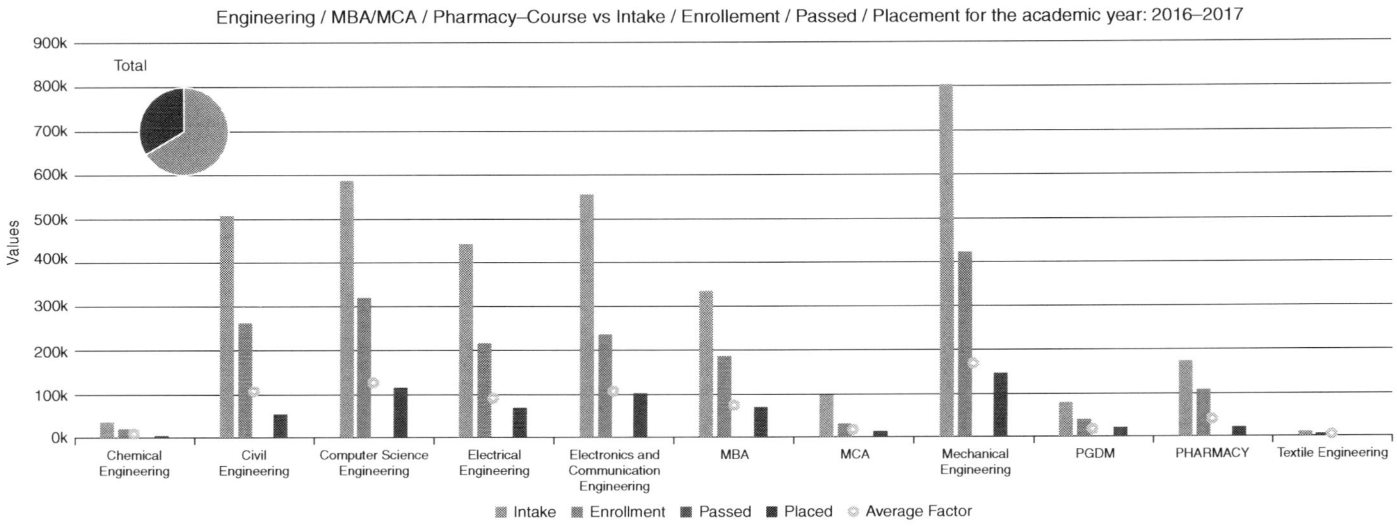

FIGURE I.1 Engineering/MBA/MCA/Pharmacy-Course vs Intake/Enrollment/Passed/Placement for the academic year 2016–2017
Source: All India Council for Technical Education (www.aicte-india.org/)

led to a burst in its popularity is a democratization of access to big data, cloud computing and AI tools. Companies such as Yahoo, Twitter and Amazon are developing tools and making it open source. These tools are freely available for anyone to access, use and develop further.

Thinking of what might transpire in AI in the near future, one just has to look at research projects being pursued by the labs in Silicon Valley. Real breakthroughs in the interface of AI and materials science, quantum computing and probabilistic programming might potentially happen in the not-so-distant future, and that possibility looks extremely realistic. Quantum computation, if it can go beyond the simplistic problems that it solves currently, can really take AI research to an entirely new plane. It can usher in a new era of even bigger data and computing. The new forms of AI that follow can barely even be conceived of now. For example, Artificial General Intelligence (AGI) might finally occur and that will be the pinnacle of success for research efforts in AI.

AGI stands for the intelligence of a machine, presumably developed in the future, that can perform multiple tasks as performed by humans. AGI is also referred to as "strong AI" or "full AI". To put it simply, it is one single machine that can drive your car, compose music, strike conversation, react intelligently in emergency situations and can do everything that can be done by a human being. It is essentially the ability of a machine to perform "general intelligent tasks". In a layperson's language, there is a distinction between strong AI and "weak AI", which simply is the ability to accomplish a specific problem-solving or reasoning task. Weak AI, in contrast to strong AI, does not attempt to perform the full range of human cognitive abilities.

An even more futuristic idea is that of Superintelligence. As AI gets better over time, it is conceivable that at some point in the future it will start improving itself, just like evolution of *Homosapiens*, only infinitely faster than a natural process over the course of history. Humans will not be required any longer to develop the next generation of AI. Once AI has the ability to improve itself, the previous version of AI will be able to further improve itself and develop a better, next version. In other words, it will iteratively self-improve. This runaway self-improvement might lead to superintelligence – a form of intelligence that is fantastically better than humans in all aspects. There will be no reason for this advanced form of AI to stop self-improvement. The result will be a runaway explosion of intelligence, called the singularity, that experts conjecture will change human history forever and in an irreversible manner.

Where will all this leave human beings? Given the promises as well as the challenges from automation, robots and AI, where will humans be in another 20, 50, or a hundred years? Will humans be only mute spectators as robots become increasingly commonplace? Will there be widespread resentment as humans are replaced by AI? Will the issue of AI increasingly polarize people, like climate change issues? In this day and age of fake news and post truths, is there a danger of misinformation spreading about AI? Or, will the exact opposite happen and automation and AI only evolve in a direction that is beneficial to mankind? Perhaps new forms of AI will better serve humans in medicine, transport, risk management, education, etc.

Perhaps, certain forms of AI will make the judicial system better. AI in workplaces will prevent labor frustration from repetitive chores. Which way will AI really develop – in a direction that is positive for humans or negative?

George Friedman in his book, *The Next 100 Years: A Forecast for the 21st Century*, writes "Anger does not make history.[6] Power does. And power may be supplemented by anger, but it derives from more fundamental realities; geography, demographics, technology, and culture". Similarly, fearing it or getting angry with AI will not help human beings. One needs understanding of AI to deal with it and be prepared for it. This means developing a comprehensive and consistent response. And, no one can do it alone.

One needs all the stakeholders, including countries, governments, private corporations, academicians, philosophers and AI researchers, to come together. The attempt should be to listen to anyone who is impacted by AI, and hence this will also include all direct and indirect stakeholders, for example, developers, engineers, civil society, standard-setting bodies, and even labor unions, medical ethics review bodies and AI practitioners. All of them collectively need to think about how to proceed on AI, how to develop a code of ethics for AI and how to ensure that AI remains just like other technologies, always under the control of humans. This is not going to be easy but unfortunately that is the only option, and unless there is a collective effort to formulate a policy on AI, it will be an understatement to say that humankind is playing with fire.

Most importantly, AI needs to become the topic of mainstream discussion in the media and among the general public. Whether AI is a friend or an enemy, the fact remains that it is faceless and nameless and doesn't excite the same visceral reaction that an illegal immigrant or outsourcing to other countries does among politicians and public alike. Unless the public understands the real issues and starts getting involved and the media diligently works on shaping an informed public opinion on AI, policy makers will not be in a hurry to address the core issues and resolve the conflicts that arise out of the development and use of AI.

In forming a sensible, balanced and forward-looking view on AI, the role of both private sector and public sector/government is equally important. The private sector needs to educate people about a) what AI has achieved so far and the likely trajectory without sensationalizing the developments; b) how they intend to use it and what it would mean for the impact on people. The public sector and government needs to talk about: a) the current level of understanding and comfort levels on what and how much they are going to allow in terms of proposed regulation; b) how they are going to handle the challenges and assuring people that they will be proactive; c) how they will coordinate with private sector and other stakeholders in arriving at a comprehensive response plan.

In a democracy, the politicians usually take up and address issues on the basis of their electoral potential, based on whether these issues strike a chord with the public. Will people understand the issue? Will it be able to capture the

TABLE I.1 Top myths of AI

	Myth	*Fact*
1	Superintelligence by 2100 is impossible.	It may happen in decades, centuries or never.
2	Superintelligence by 2100 is inevitable.	Experts disagree and we simply don't know.
3	Only Luddites worry about AI.	Many Top AI researchers are concerned.
4	AI turning evil or conscious is a worry.	AI turning competent with goals misaligned with ours is a worry.
5	Robots are the main concern.	Misaligned intelligence is the main concern.
6	AI can't control humans.	Intelligence enables control; we control animals by being smarter.
7	Machines can't have goals.	A heat-seeking missile has a goal.
8	Superintelligence is just years away.	It's at least decades away, but it may take that long to make it safe.

Source: Based on "Benefits & risks of artificial intelligence", The Future of Life Institute. https://futureoflife.org/background/benefits-risks-of-artificial-intelligence/

imagination of the general population? Will there be an emotional reaction from these debates and will politicians be able to find any scapegoat who can be blamed? These are some of the factors which determine the fate of an issue from a politician's perspective and whether they take it up. These political considerations also apply to AI as a public issue.

The top myths of AI

There is a lot of confusion about the future of AI and what it will/should mean for humanity, and there are several issues where experts disagree. Hence, it is important to look at some of the most common myths related to AI.[7]

Understanding and debate is the only way forward

Thankfully, a small but steadily increasing number of influential people have publicly started voicing concerns regarding AI. The likes of Elon Musk, Bill Gates, Stephen Hawking, Mark Zuckerberg and Nick Bostrom have voiced concern about what AI will do to human society. While a few years ago any such concerns would have been considered premature at best and outright nonsense at worst, there is at least a sense now that such concerns might have some basis. In addition, there have been some steps in the right direction already. In January 2017, a number of experts from a wide range of backgrounds assembled together in Asilomar, California to formulate the first set of Principles to guide the future direction of AI research and development.

There have been important developments in the direction of making AI more responsible and to enhance the positive impact while mitigating the negative consequences. With Amazon, Apple, DeepMind, Facebook, Google, IBM, and Microsoft as its founding partners and several for-profits and non-profits as partners, *Partnership on AI*[8] is an effort to ensure that we can invest more attention and effort in harnessing AI to contribute to solutions to some of humanity's most challenging problems, including making advances in health and well-being, transportation, education, and the sciences. Another important development was *The Forum on the Socially Responsible Development of Artificial Intelligence*,[9] held in Montreal in November 2017, which concluded with the unveiling of the preamble of a draft declaration to which the public is invited to contribute in a co-construction process involving all sectors of society. The Montreal convention brought together about 400 participants around the themes of cybersecurity, legal liability, moral psychology, the job market, the health system and the concept of "smart cities".

We firmly believe that the technology of the future, and that includes automation, machine learning and AI, will increasingly have a bigger and bigger impact on human society. There are a number of forces at work that make this shift towards AI irreversible. We also believe that the biggest mistake we could make right now is to ignore the developments and avoid discussions regarding what the long-term impact of AI might be. Unlike every other thing we have seen so far, AI both promises and threatens to be one of the most significant developments in all of human history.

AI has the potential to bring massive changes to the future of mankind and it is for us to decide if these changes will be good or bad. We write this book with the hope that it will motivate the reader to find out more about these issues, and help ignite an informed and knowledgeable discussion on the subject. As Albert Einstein[10] has said ***"The important thing is not to stop questioning. Curiosity has its own reason for existence."***[11]

Notes

1 Eliezer Yudkowsky is an AI researcher, blogger and exponent of human rationality (specifically, his Bayes-based version of it). Yudkowsky cofounded and works at the Machine Intelligence Research Institute (formerly the Singularity Institute for Artificial Intelligence), a non-profit organization that concerns itself with the concept known as the singularity. Yudkowsky also founded the blog community *LessWrong* as a sister site and offshoot of *Overcoming Bias*, where he began his blogging career with economist Robin Hanson. https://rationalwiki.org/wiki/Eliezer_Yudkowsky (Accessed on November 21, 2017)

2 Warren Edward Buffett is an American business magnate, investor and philanthropist. Buffett serves as the Chief Executive Officer and Chairman of Berkshire Hathaway. He is considered by some to be one of the most successful investors in the world, and he is also one of the wealthiest in the world. https://en.wikipedia.org/wiki/Warren_Buffett (Accessed on November 21, 2017)

3 William Henry Gates III, or more popularly, Bill Gates, is an American businessman and philanthropist. One of the richest persons in the world, Bill Gates is the co-founder of Microsoft, one of the world's largest technology companies. Through his foundation, he, along with his wife, Melinda, has been very active in doing philanthropic work. Though he has remained associated with Microsoft as an owner, he has delegated the executive responsibilities to professional managers. https://en.wikipedia.org/wiki/Bill_Gates (Accessed on November 15, 2017)
4 www.linkedin.com/pulse/warren-buffett-predicts-significantly-less-employment-chip-cutter/ (Accessed on November 21, 2017)
5 All India Council for Technical Education (www.aicte-india.org/).
6 George Friedman is a US geopolitical forecaster and strategist on international affairs. He is the founder and chairman of *Geopolitical Futures*, a new online publication that analyzes the course of global events. Prior to founding *Geopolitical Futures*, Friedman was chairman of Stratfor, the private intelligence publishing and consulting firm he founded in 1996. Friedman resigned from Stratfor in 2015. https://en.wikipedia.org/wiki/George_Friedman (Accessed on November 21, 2017)
7 "Everything we love about civilization is a product of intelligence, so amplifying our human intelligence with artificial intelligence has the potential of helping civilization flourish like never before – as long as we manage to keep the technology beneficial" – Max Tegmark, President of the Future of Life Institute, "Benefits & risks of artificial intelligence". https://futureoflife.org/background/benefits-risks-of-artificial-intelligence/ (Accessed on February 14, 2018)
8 Partnership on AI (full name Partnership on Artificial Intelligence to Benefit People and Society) is a technology industry consortium focused on establishing best practices for AI systems and to educate the public about AI. Publicly announced in September 28, 2016, its founding members are Amazon, Facebook, Google, DeepMind, Microsoft, and IBM, with interim co-chairs Eric Horvitz of Microsoft Research and Mustafa Suleyman of DeepMind. Apple joined the consortium as a founding member in January 2017. In January 2017, Apple head of advanced development for Siri, Tom Gruber, joined the Partnership on AI's board. https://en.wikipedia.org/wiki/Partnership_on_AI (Accessed on February 14, 2018), www.partnershiponai.org/ (Accessed on February 14, 2018)
9 AI can eliminate mundane jobs, but it could also lead to widespread unemployment. The Montreal Forum on the Socially Responsible Development of Artificial Intelligence was organized in November 2017 and resulted in *The Montreal Declaration for a Responsible Development of Artificial Intelligence: a participatory process* (3 November 2017) http://nouvelles.umontreal.ca/en/article/2017/11/03/montreal-declaration-for-a-responsible-development-of-artificial-intelligence/ (Accessed on February 14, 2018), http://montrealgazette.com/business/responsible-ai-conference (Accessed on February 14, 2018)
10 Albert Einstein was a German-born theoretical physicist. Einstein developed the theory of relativity, one of the two pillars of modern physics (alongside quantum mechanics). Einstein's work is also known for its influence on the philosophy of science. Einstein is best known by the general public for his mass–energy equivalence formula $E = mc^2$ (which has been dubbed "the world's most famous equation"). He received the 1921 Nobel Prize in Physics "for his services to theoretical physics, and especially for his discovery of the law of the photoelectric effect", a pivotal step in the evolution of quantum theory. https://en.wikipedia.org/wiki/Albert_Einstein (Accessed on November 21, 2017)
11 https://books.google.co.in/books?id=dlYEAAAAMBAJ&lpg=PA61&pg=PA64#v=onepage&q&f=false (Accessed on April 17, 2018)

1

AI—THE HISTORY AND EVOLUTION

> I believe that at the end of the century the use of words and general educated opinion will have altered so much that one will be able to speak of machines thinking without expecting to be contradicted.
>
> —*Alan Turing*[1]

Intelligence: There is still no agreement on its definition

The definition of intelligence is highly subjective and there is hardly a consensus over what intelligence really means. It can be defined as the capacity for logic, understanding, self-awareness, learning, emotional knowledge, planning, creativity and problem solving. It can be more generally described as the ability or inclination to perceive or deduce information, and to retain it as knowledge to be applied towards adaptive behaviors within an environment.[2]

Merriam-Webster defines intelligence as the ability to learn or understand things or to deal with new or difficult situations.[3] Albert Einstein once said, "The true sign of intelligence is not knowledge but imagination." Socrates said, "I know that I am intelligent, because I know that I know nothing." For centuries, people have tried to define intelligence.[4] The following passage on intelligence is from "Mainstream Science on Intelligence" (1994), an op-ed statement in the *Wall Street Journal* signed by 52 researchers:[5]

> A very general mental capability that, among other things, involves the ability to reason, plan, solve problems, think abstractly, comprehend complex ideas, learn quickly and learn from experience. It is not merely book learning, a narrow academic skill, or test-taking smarts. Rather, it reflects a

> broader and deeper capability for comprehending our surroundings—"catching on," "making sense" of things, or "figuring out" what to do.

The American Psychological Association[6] notes that intelligence refers to intellectual functioning. Intelligence quotients, or IQ tests, compare your performance with other people your age who take the same test. These tests don't measure all kinds of intelligence, however. For example, such tests can't identify differences in social intelligence, the expertise people bring to their interactions with others. There are also generational differences in the population as a whole. Better nutrition, more education and other factors have resulted in IQ improvements for each generation.[7]

Natural intelligence originates in biology. Simply speaking, natural intelligence is how animal or human brains function. But, the definition of intelligence which links it to the brain alone is a narrow one as nature also demonstrates non-neural control in plants and protozoa and distributed intelligence in colony species like ants, hyenas and humans.[8] However, for all practical purposes, *human intelligence* is referred to as natural intelligence.

Natural intelligence vs. artificially created intelligence

Human intelligence or natural intelligence is created naturally and biologically. This is different from intelligence which is created by humans in machines using technology. It is expected that there will be differences between natural intelligence and artificially created intelligence.[9] Naturally, AI has significant dominance in many tasks, especially when it comes to monotonous judgments. In contrast, biological neural networks also have superiority in some qualities:

1. **Speed** – Natural intelligence will be at a disadvantage versus an artificially created intelligence and this is similar to the advantage a computer has over the human brain in performing calculations. Natural intelligence will also have more "down time".
2. **Objectivity** – Humans have inherent biases in their decision-making and in most cases these biases can be eliminated by using an artificially created intelligence. The artificially created intelligence can also be more accurate and precise.
3. **Handling complex and different tasks** – Machines are usually designed for a few tasks while humans can do better at handling more complex and different types of activities. Humans will also have an advantage in multitasking.
4. **Complex movements** – Even the most advanced robots can hardly compete in mobility with humans. This makes humans more suitable for tasks that require a higher degree of maneuverability.

5. **Evolution** – The pace of evolution in humans is very slow and it takes thousands of years to make the changes and modifications required as per the change in the environment. But machines can adapt very quickly.
6. **Cost** –Natural intelligence resides in humans and the life cycle cost of humans has not changed much over thousands of years when we talk about covering the basic necessities. However, the cost of creating, operating and maintaining machines is reducing at a fast pace.

The beginning of AI

Ideas and fantasies about machines that can think like humans have been around for a long time. One of the primary drivers of this thought process was the desire to create something that had all the human abilities but none of the human shortcomings. The idea was to create a machine which can think and act like a human even in subjective and ambiguous situations, but which does not get interrupted by mundane distractions and the machine also always remains consistent, unlike humans. It is something that doesn't feel hunger, doesn't get angry, doesn't get tired and is not prone to making mistakes.

Science-fiction writers, unshackled by constraints of reality and not bound by the rules of contemporary science, have thought of thinking machines for a very long time. These thinking machines were characterized in various ways, but one of them was to call them *intelligent*. Not in a human sense, but in an *artificial* manner. These machines would then be called "artificially intelligent" or considered to have "artificial intelligence".

The idea behind AI is different from that of a mere computational device. A computer, as a computation machine, divides each and every task into more manageable, smaller tasks by representing it in the form of "0s" and "1s", which effectively means that anything which can be expressed in "black and white" can be handled by a computer. The idea behind AI is that of a machine which is more human-like, a machine which can handle ambiguous situations as well as humans can, and/or which can also look at situations and possible outcomes in a probabilistic manner to handle the "grey" areas.

Any entity, whether it is a machine, an algorithm or a robot which can think and take decisions independently on objective and subjective matters, without the guidance of humans, can be said to have AI. Anything that doesn't reside in the human body or has not developed as a result of a natural evolutionary biological process over millions of years, and was developed in laboratories by scientists and engineers, would qualify as AI. Under this definition, AI can function like humans in tasks where the directives are not always straightforward and mathematically definable. Furthermore, the development of cognitive artificial systems which are able to perform jobs typically executed by humans, such as two-way conversation or instant decision-making, are all part of the development of AI.

Many of the developments which brought us to the current state of affairs in AI are relatively recent. It would be pertinent to highlight the role of World War II (WWII) in this. The all-around scientific development during WWII also included a lot of interest in computation and statistical thinking from the political and military leaders. For example, the interception of secret communication code and code-breaking was a big war effort on both sides, and there is sufficient evidence available of instances where the Allied Powers used it very effectively to gain an upper hand over Axis Powers such as Germany.[10]

Where all of this began less than 100 years ago: Rossum's robots

The word "robot" can be traced back to a 1920 play, *R.U.R.* or *Rossum's Universal Robots,*[11] by the Czech writer Karel Capek. In Czech, *robota* means serfdom. Rossum's robots were synthetically produced, servile, flesh and blood creatures. The robots willingly agree to direction from humans and gradually replace humans in various activities. Robots become cheap, are readily available everywhere, and become indispensable as a source of cheap labor. As the play progresses, however, Rossum's robots become discontent, stage a revolution, and kill all humans. The play ends with some hope though as a few robots start displaying human-like emotions, and there is some hope that they might evolve into a human-like form. The play itself would have been forgotten, but has become immortal because of the word *robot* it contributed.

Isaac Asimov coined the term *robotics* in the 1940s while writing his hugely popular robot stories.[12] Asimov's Three Laws of Robotics, serving as the basis of robot-ethics, became well known among science-fiction readers. As introduced in Asimov's 1942 short story, "Runaround", the three laws read: (1) A robot may not injure a human being or, through inaction, allow a human being to come to harm; (2) A robot must obey the orders given to it by human beings except where such orders would conflict with the First Law; (3) A robot must protect its own existence as long as such protection does not conflict with the First or Second Laws. These laws provided a framework for the action of robots in Asimov's stories, and are so natural that they were referenced by other authors and even in other genres.

Some of these fictional ideas assumed a semblance of reality as the British mathematician and the famous WWII code breaker, Alan Turing,[13] wrote a paper in 1950 in the journal *Mind* which introduced the Imitation Game[13]. In his paper, Turing developed a method to find out if a machine can exhibit intelligent behavior indistinguishable from that of a human. Turing really wanted to investigate if a machine could, in principle, think. The paper starts with the words, "I propose to consider the question, 'Can machines think?'".

However, steering clear of the ambiguity that the concepts of thought and machine could lead to, Turing instead devised a more practical test—the Imitation Game—through which an interrogator could determine whether the responder was a human or a machine. The machine's job, in this game, is to try to fool the interrogator into thinking that it is a human. Turing suggested that the machine could convince a human about 30 percent of the time in a five-minute conversation. Turing's paper did not have much of a role in the practical development of AI. However, the philosophical impact of the idea was significant; one could not dismiss the idea of thinking machines so easily.

Turing did not think of his paper in a vacuum. He was a member of the Ratio Club,[14] a group of young upcoming but really smart mathematicians, engineers, psychologists and physiologists, who used to meet regularly to discuss issues in cybernetics. Favoring the young and the daring, professors were banned from the group. Any member attaining professorship in his professional academic life was expected to immediately resign from the group. The group discussions tended to be extremely thought provoking. Many of the members went on to contribute significantly to their fields.

Among the members was Horace Barlow, Charles Darwin's great-grandson, who made significant contributions to the neuroscience of vision. Albert Uttley, another member of the group, researched radar during WWII. Thomas Gold, an astrophysicist who helped solve the mystery of pulsar, a type of rotating neutron star, was also a member of the club. Machine intelligence was a frequent topic of discussion at the group meetings, and it is very possible that Turing was motivated to write his paper following discussions with the Ratio Club members.

The Dartmouth workshop: Assembly of the finest minds

A number of developments after WWII led to an interest in automated and intelligent systems. Perhaps, if one were asked to name any one single event that was crucial in the development of AI, it might be the **Dartmouth workshop** held in the summer of 1956.[15] The idea for the workshop was conceived by John McCarthy,[16] a young mathematics professor at Dartmouth, and was further developed by Marvin Minsky,[17] Nathaniel Rochester[18] and Claude Shannon.[19] John McCarthy came up with the name Artificial Intelligence and the workshop was called the Dartmouth Summer Research Project on Artificial Intelligence.

A number of mathematicians from universities and leading research centers, such as IBM, attended the workshop. The attendees also included John Nash,[20] a pioneer in game theory and a future Nobel Prize Winner who is also the subject of a popular and successful Hollywood movie, *The Beautiful Mind*. The workshop was important not so much in providing immediate solutions to AI problems, as it was in defining this new subject and the potential problems, opportunities, implications and practical applications that were relevant for the field.

The proposal for the workshop listed a number of topics which, as years passed by, developed into areas of significant AI research and pushed the frontiers for this nascent but exciting stream of science and computing.[21] The topics included: Automatic Computers, How Can a Computer be Programmed to Use a

Language, Neuron Nets, Theory of the Size of a Calculation, Self-Improvement (of Intelligent Machines), Abstractions, Randomness and Creativity. It is remarkable that a number of areas whose breakthroughs have brought us to the current stage of AI development were already being defined and discussed at this workshop.

In any scientific development, defining the basic structure and the problems is the essential key to progress in that area. The Dartmouth workshop was instrumental in bringing together a group of like-minded thinkers who systematically defined the problems in AI. According to all accounts, the participants came out energized and motivated to follow the program. Other similar workshops soon followed. The Teddington Conference on the Mechanization of Thought Processes was held in the UK in 1958, attended by some of the same Dartmouth workshop participants, including John McCarthy and Marvin Minsky. ***AI as a research field was born and was vibrant...***

1960s

The 1960s saw gradual research development in many areas. Programs started in multiple labs to develop chess and checkers playing machines. Fortran[22] and LISP,[23] two important computer languages, became popular and went on to see wide usage. Natural language processing,[24] study of the interaction of AI and human languages, also saw a lot of activity. The world's first robotics company, Unimation,[25] was founded in 1962. Unimation supplied robotic arms to car manufacturers, which led to the automation of tasks at scale. But as usually happens with any other scientific field, there were also setbacks. Like any young fledgling field with big promises, the field of AI saw skeptics too. A 1966 report on machine translation by the Automatic Language Processing Advisory Committee was so negative that it effectively brought research in NLP to a halt.[26] Similarly, an influential paper by Marvin Minsky and Seymour Papert expressed much skepticism about neural nets and dissuaded any research in the field for an entire generation.[27]

1970–80s

The decades of the 1970s and 1980s saw a massive increase in the computing capacity of machines. As computers became more ubiquitous with cheaper cost, less complicated handling, ease of maintenance and a variety of new practical applications, this was a big boon for the development of hardware for AI. Computer science departments across multiple universities aggressively recruited brilliant and talented students and faculty members. With talent being attracted to the world of computing, there was swift progress in computer programs and the development of new and better algorithms. There was also enhanced public perception of computers with the emergence of the PC. In the corporate world, companies also started to see immense commercial scope in this field and started their own research and development (R&D). However, from the perspective of AI, the 1970s and 80s saw gradual progress, and remained confined to research laboratories.

1990s

In the 1990s, more public instances of AI versus human competition took place. Researchers at IBM created TD-Gammon,[28] a computer backgammon program based on neural nets, which could play at a proficiency level close to the top backgammon players. In 1996, Deep Blue of IBM made history by being the first computer program to beat a reigning world champion in chess, Gary Kasparov.[29] The cultural impact of Deep Blue was big. In popular perception, a machine beating the world champion in one of the more cerebral games in the history of humankind was nothing short of a revolution.

The World Wide Web gradually came into widespread use.[30] This was not only helpful for easy access of information for the masses; it also helped with faster dissemination of research around the world to the interested scholars and scientists. The advent and quick spread of the internet also convinced the computer scientists and researchers that the applications of computing could be extremely varied and useful.

While the 1900s saw major advances in AI and computing, the most important advances, responsible for making AI what it is today, took place in the 21st century as big data, cloud computing, and a number of computational methods such as neural nets and bayesian methods came together. As we discuss in several other places in this book, AI as a technological innovation depended on many enablers for its progress; several advancements played a role and continue to contribute to its development.

Scientific development is rarely linear: The AI winters

Scientific progress has bumps and speed breakers and things never move in a straight line; this was the case with AI too. Just like other research areas, it was not all smooth sailing for AI research. There were two separate epochs, the first during the period 1974–1980, and then from 1987 to 1993, which saw a dramatic decline in AI research and funding. These two periods are now known as the first and the second AI winters.

While a number of reasons led to the onset of the first AI winter, one can ascribe special importance to the influential book, *Perceptrons*,[31] by Marvin Minsky and Seymour Papert. The authors conducted a detailed analysis of the early versions of neural nets, and pointed out the major limitations of these networks. Minsky and Papert were influential figures and their book single-handedly dissuaded a generation of researchers from pursuing neural nets. When the potential of further research on neural nets was seen as limited, it negatively affected the interest levels of researchers.

Another reason for the AI winter was simply the chasm between the aspirations of AI ideas, and the difficulty of implementation. Data was still stored in analog devices, putting a practical limit on data storage and retrieval. Many human characteristics, such as speech and vision, proved to be incredibly difficult problems for AI to crack. Progress in these fields had to wait for developments to occur in other technological fields. Agencies that funded AI research, such as the British government and DARPA in the US, became frustrated with the lack of progress and reduced available funding.

The second AI winter,[32] from 1987 to 1993, happened as desktop computers, such as those made by Apple and IBM, quickly gained in popularity and made previous specialized AI hardware obsolete. Expensive LISP machines and expert systems became irrelevant. Before people realized, a half-billion-dollar industry collapsed overnight. In the early 1980s, realizing the importance of AI-related research; the US government had formed the Strategic Computing Initiative to fund research in this field. The idea was to fund a whole spectrum of projects, big and small, whose results would be brought together to achieve the big goals of AI. However, as the enthusiasm for the field waned by 1988, the Strategic Computing Initiative decided to cancel funding for AI research.[33]

The 21st century: Rise of big data, cloud computing and AI-ML tools

While Deep Blue beating Kasparov at chess was a significant achievement, skeptics still expressed doubts at how fast AI could progress. The general opinion was that it would be a long time before AI could beat humans at more complex games such as *Go*, or *Texas Hold'Em*. After all, could AI really bluff at a game of poker? But progress has been made at a fast and furious pace. In 2016–17, the AI *Go* player, named AlphaGo, played around 60 games with the top *Go* players and won every single one of these.[34] In January 2017, another milestone was achieved, when an AI program, named Libratus, showed it could bluff like the pros and was able to defeat top players at *Texas Hold'Em*.[35]

A variety of scientific and technological developments have come together in the new millennium to give AI research a major boost. It has become easier to store an ever-increasing amount of data easily and cheaply. Computational resources have become powerful enough to enable this data to be sieved through easily. Over the last 20 years, our ability to store, analyze and manipulate tremendous amounts of data has increased by leaps and bounds.

The need to own and pay for expensive infrastructure to do AI research has decreased considerably with the advent of cloud computing. And a number of mathematical and statistical tools in machine learning have become widely tested and freely available. Access to big data, cloud computing and AI-ML (Artificial

Intelligence-Machine Learning) has also become democratized, with researchers and companies willing to share information. It is possible, affordable and practical now for a high school student to open an account on the Amazon or Google websites and start doing AI-related tinkering. The easy accessibility of big data methods, cloud computing and computational methods could potentially make progress even faster.

Big data: Explosion of data in the new millennium

In the last century, almost all information was stored in analog containers such as books, disks, magnetic tapes, etc. At the turn of the new millennium, when the world was fixated on solving the Y2K problem, digital storage swiftly replaced analog devices as the preferred medium for information storage. The world's capacity to store information has increased drastically over the last 20 years, exploding from a mere 2 billion gigabytes in the mid-1980s, to over 300 billion gigabytes in late 2007.[36]

Extrapolating this trend from 2007 to 2017 would give a figure of around a trillion gigabytes in 2017. The current storage capacity is likely even more given the massive information explosion that has accompanied social media. This is a stunning amount of data, but what is impressive is how easily accessible it is to anyone, even through just a smartphone. What sets our present time apart from previous eras is the ease with which a lay person, not just experts, can access big data.

The cost of storing data has come down tremendously too. The cost of storage of 1GB of data was about $438K in 1980, $11K in 1990, and just $11 in 2000. In 2014 this cost was just 3 cents. As of 2017, Google was offering free storage for up to 15GB of data.

Computational power

An increase in data storage capability would not mean much unless accompanied by an increase in the computational power to manipulate the data. Indeed, there has been about a trillion-fold increase in the processing power of computers from the 1950s to 2017. To understand how much computing power is at a lay person's fingertips, consider the fact that a single Apple iPhone 5 has 2.7 times the processing power of the 1985 Cray-2 supercomputer—a thought which would have been laughable in the mid-1980s. Another example showing the speed of development in computation is that the Samsung Galaxy S6 is the equivalent of five PlayStation 2s in terms of processing power. Computing power has not only exponentially increased; it has become easily available to ordinary people.[37]

Cloud computing

The final development leading to the democratization of data and computation is the evolution of cloud-based computing. With the advent of the cloud, infrastructure can be delivered as a service. There is no need to own the complex infrastructure required to store big data and slice and dice it. One can access data, as well as the tools to analyze that data, by accessing the cloud. It has become possible for a reasonably smart school student to open an Amazon or Google cloud service account, and do serious analysis on it. It has also become possible for small start-ups to very quickly start exploring projects without the need to set up the computing infrastructure and employ personnel to maintain that infrastructure. This has allowed for new developments to occur at an even faster rate.

The impact of big data, computing power and cloud computing

Big data, computing power and cloud computing have all been important in making huge amounts of data and enormous computing power available to anyone who is interested. But for AI to develop, what has also been important is the science and mathematics of AI. In the decades following the Dartmouth workshop, a number of mathematical learning techniques developed that could help a machine behave intelligently. All of these techniques—Neural Nets, Support Vector Machines, Bayesian Methods, to name a few—went through bouts of popularity and abandonment. One method, however, which has received a remarkable revival in interest, is Connectionism.

Thinking like a child

Of the many approaches to AI, one class of methods called connectionism stands out as particularly promising.[38] The idea is motivated by studies of the brain. The basic building block of the brain is the neuron. Small networks of neurons, each set responsible for a different task, connect with each other to give rise to the collective phenomena of the mind. While each small neural network is responsible for a simple task, the networks come together to work as a whole. The whole is much more complex and richer and able than just a simple sum of its individual parts. Connectionism builds on this idea. The classic analogy of network effect is applicable here. As the number of neurons (or nodes) and the interconnection among them increases, the brain (or the machine) becomes more and more powerful and complex.

The early connectionism approaches were based on an electronic device called perceptron. Frank Rosenblatt,[39] a psychology professor at Cornell University, was studying ways to artificially replicate the behavior of the human brain. Rosenblatt's perceptron was a simplified electronic imitation of neuron cells.[40] Perceptrons were combined to make neural nets, networks of perceptrons, which could perform basic pattern-recognition tasks. Perceptron could be trained to recognize different types of objects, and then it could be made to try to predict the type of new objects. Rosenblatt and his perceptron received quite a bit of popular attention. However, as much promise as the perceptron displayed, it also received a number of criticisms. As we mentioned earlier, Marvin Minsky and Seymour Papert wrote an influential book in 1969 discussing the limitations of the perceptron. This book, in particular, dissuaded researchers from further studying and developing neural networks.

Interest reemerged in neural networks in the 1990s as researchers found ways to circumvent objections raised by Minsky and Papert. Ideas from other fields, such as physics and biology, also helped the researchers find new variations of the original perceptron-based networks. One of the current popular AI buzzwords, Deep Learning, is an overarching rubric for connectionism-inspired ideas and techniques. Deep learning has shown immense potential in a variety of areas which require complex learning followed by action. Deep learning tools can learn from training data, and then act on new incoming data.

In many ways, deep learning is similar to how a child learns: through trial and error. Rarely does a child simply follow the instructions of his or her parents when it comes to the dos and the don'ts. A toddler does not just see a demonstration once, and master it free of errors. A child's habits and temperament slowly take shape as he or she experiences the world around him or her. If an experience is pleasant, the child repeats it. If an experience is painful, he or she runs away from it. Deep learning explores and learns in a similar fashion. It has an pre-programmed sense of what its goal is. Then, through experience, making both correct and wrong choices, it learns to deliver. If it is a good learner, it starts delivering results in an efficient way quickly. Many of the new developments in AI—driverless cars, face recognition technology, Siri, IBM Watson—use deep learning in one form or another.

AI can be a source of BIG positive change in the world

AI has started to attract attention not just for its different characteristics vs. other technologies but also for its practical applications. In her article[41] for *Business Insider*, Kriti Sharma, Vice President of AI at Sage Group, wrote that adopting technologies like AI can make business more productive by cutting down the time people need to spend doing basic administrative tasks. For example, a technology like chatbot—especially as the first line of customer interaction before speaking to a person over the phone—can cut down customer wait times significantly.

The article further states that AI won't harm businesses; on the contrary, it will make them more productive. Businesses could grow faster if more time could be devoted to real value-added activities for customers and industries—like improving strategy, creating better products or spending more time with customers. Adopting AI can be cost-effective, complementary to customer engagement and useful in closing talent gaps, and you don't need to become an AI expert to reap these rewards. The other aspect is that consumers too will benefit from the increasing role of AI.

AI has several advantages, including using it to reduce the margin of error of its use in the exploration of unsafe, hazardous and hostile environments. AI can also do well in repetitive and monotonous jobs or the areas where analysis of past data can help in arriving at the best possible course of action. For example, there could be numerous applications of AI for medical professionals. AI can help make doctors more efficient by providing them with the information they need in a timely and orderly manner. AI can also make surgeries more precise, less dangerous and less painful so that patients can recover faster.[42] In short, AI can help doctors in arriving at the correct diagnosis for a patient and to also recommend the best therapy.

In a European Parliament briefing, *Artificial Intelligence: Potential Benefits and Ethical Consideration*, the key finding is that the ability of AI systems to transform vast amounts of complex, ambiguous information into insight has the potential to reveal long-held secrets and help solve some of the world's most enduring problems.[43] There are many useful examples of AI in our daily life. Any credit card purchase, finding the shortest route through GPS, spam filters and recommender systems of e-commerce businesses are all AI based. There are others as well: the Google Translate service, search engine optimization algorithms and face recognition are also AI applications. Siri, the iPhone app that understands our words and responds in a sensible and useful way, is based on AI algorithms too.

The possibilities are truly amazing and endless with AI and that is why all leading companies including Amazon, Apple, Facebook, Google, IBM, Microsoft and others are investing in AI. Google is building self-driving cars and Facebook is getting bigger and bigger in AI. Apple has developed Siri; Microsoft has built Cortana, a similar personalized assistant; Google has acquired DeepMind, whose long-term aim is to build general AI; and IBM is investing a huge amount of resources in applying its Watson cognitive computing system to the medical domain, to finance, and to personalized education.

AI is unlike anything else we have seen before

Over the last approximately 60 years, the history of AI has alternated between rapid development and states of hiatus. However, things have never been as exciting as they are today. With progress in core areas such as computing capability, processing power and storage capacity, along with advances in our understanding of the human brain, AI research is progressing at a very fast pace. With all this progress, broader questions are also arising regarding AI's promises as well as threats. These questions have certainly become significant enough that discussions on AI should no longer remain confined to university campuses, research labs and unicorns-in-making any longer.

In the next chapter, *Why AI is Unlike Any Other Technology*, we will discuss how AI is fundamentally different from other technologies that humans has seen so far. The technologies which have emerged in the past and have become integral parts of our lives have aimed at making our lives more comfortable and efficient. But, the *thinking* part has always been under the control of humans. Thinking was never outsourced to machines. Is it possible that, with AI, humans will start outsourcing the thinking aspect? We will discuss such possibilities in the subsequent chapters.

Notes

1 Alan Turing (1912–1954) is considered to be the father of modern computing and AI. His concept of the Turing machine is still one of the most widely examined theories of computation. His early work was undertaken at King's College, Cambridge. From 1936 to July 1938 he studied mathematical logic at Princeton University. Turing obtained his PhD in June 1938. He has proposed an experiment now known as the Turing test, an attempt to define a standard for a machine to be called "intelligent". The idea was that a computer could be said to "think" if it could fool an interrogator into believing that the conversation was with a human. www.maths.manchester.ac.uk/about-us/history/alan-turing/ (Accessed on October 23, 2017).
2 http://courses.aiu.edu/BASIC%20PROCESSES%20OF%20THOUGHT/Sec%204/SEC%204%20BASIC.pdf (Accessed on 11 April 2018)
3 www.learnersdictionary.com/definition/intelligence (Accessed on October 23, 2017)
4 http://bigthink.com/going-mental/what-is-intelligence-2 (Accessed on October 23, 2017)
5 www.journals.elsevier.com/intelligence/, www.personalityresearch.org/journals/intelligence.html (Accessed on April 11, 2018)
6 The American Psychological Association (APA) is the leading scientific and professional organization representing psychology in the United States, with more than 115,700 researchers, educators, clinicians, consultants and students as its members. APA's mission is to advance the creation, communication and application of psychological knowledge to benefit society and improve people's lives. www.apa.org/about/index.aspx (Accessed on October 24, 2017)
7 www.apa.org/topics/intelligence/ (Accessed on October 24, 2017)
8 www.cs.bath.ac.uk/~jjb/web/uni.html (Accessed on October 23, 2017)
9 From the comments of EgorDezhic, AI and CogSci enthusiast. Author of *CognitiveChaos.com* at https://becominghuman.ai/artificial-vs-natural-intelligence-626b6c7addb2 (Accessed on October 24, 2017)
10 From September 1938, Alan Turing worked for the Government Code and Cypher School (GCCS) on the problem of the German Enigma machine. Turing designed the bombe, named after the original Polish designed *bombakryptologiczna* (or cryptologic bomb). The bombe, with an enhancement suggested by mathematician Gordon Welchman, became one of the primary tools used to attack Enigma-protected message traffic. www.maths.manchester.ac.uk/about-us/history/alan-turing/ (Accessed on October 23, 2017)
11 *R.U.R.* is a 1920 science fiction play by the Czech writer Karel Čapek. R.U.R. stands for *Rossumovi Univerzální Roboti* (Rossum's Universal Robots). However, the English phrase Rossum's Universal Robots was used as the subtitle in the Czech original. It introduced the word "robot" to the English language. The play is about artificial people, called roboti (robots), who can think for themselves.
12 Isaac Asimov was a Jewish-American writer and professor of biochemistry at Boston University. He was known for his works of science fiction and popular science. Asimov was a prolific writer, and wrote or edited more than 500 books and an estimated 90,000 letters and postcards.

13 A. M. Turing, 1950. "Computing machinery and intelligence", *Mind*, 49: 433–60.
14 http://mitpress.universitypressscholarship.com/view/10.7551/mitpress/9780262083775.001.0001/upso-9780262083775-chapter-6 (Accessed on April 11, 2018)
15 www-formal.stanford.edu/jmc/slides/dartmouth/dartmouth/node1.html (Accessed on April 11, 2018)
16 John McCarthy was an American computer scientist and cognitive scientist. McCarthy was one of the founders of the discipline of artificial intelligence. He coined the term "artificial intelligence" (AI). http://jmc.stanford.edu/ (Accessed on April 11, 2018), https://cs.stanford.edu/memoriam/professor-john-mccarthy (Accessed on April 11, 2018), www.computerhistory.org/fellowawards/hall/john-mccarthy/ (Accessed on April 11, 2018)
17 Marvin Lee Minsky (1927–2016) was an American cognitive scientist and co-founder of the Massachusetts Institute of Technology's AI laboratory, and author of several texts concerning AI and philosophy. http://news.mit.edu/2016/marvin-minsky-obituary-0125 (Accessed on April 11, 2018)
18 Nathaniel Rochester (1919–2001) designed the IBM 701, wrote the first assembler and participated in the founding of the field of AI. At MIT, he worked in the Radiation Laboratory for three years and then moved to Sylvania Electric Products, where he was responsible for the design and construction of radar sets and other military equipment. http://history.computer.org/pioneers/rochester.html (Accessed on April 11, 2018)
19 Claude Elwood Shannon (1916–2001) was an American mathematician, electrical engineer and cryptographer known as "the father of information theory". Shannon is noted for having founded information theory with a landmark paper, "A mathematical theory of communication", which he published in 1948. www.nyu.edu/pages/linguistics/courses/v610003/shan.html (Accessed on April 11, 2018)
20 John Forbes Nash Jr. (1928–2015) was an American mathematician who made fundamental contributions to game theory, differential geometry, and the study of partial differential equations. Nash's work has provided insight into the factors that govern chance and decision-making inside complex systems found in everyday life. He shared the 1994 Nobel Memorial Prize in Economic Sciences. www.princeton.edu/news/2015/05/27/tragic-meaningful-life-legendary-princeton-mathematician-john-nash-dies (Accessed on October 26, 2017)
21 J. McCarthy, M. Minsky, N. Rochester, and C. E. Shannon, 1955. *A Proposal for the Dartmouth Summer Research Project on Artificial Intelligence*. http://raysolomonoff.com/dartmouth/boxa/dart564props.pdf (Accessed on October 26, 2017)
22 www.thocp.net/software/languages/fortran.htm (Accessed on April 11, 2018)
23 www.britannica.com/technology/LISP-computer-language (Accessed on April 11, 2018)
24 https://machinelearningmastery.com/natural-language-processing/ (Accessed on April 11, 2018)
25 Unimation was the world's first robotics company. It was founded in 1962 by Joseph F. Engelberger and George Devol and was located in Danbury, Connecticut. www.botmag.com/the-rise-and-fall-of-unimation-inc-story-of-robotics-innovation-triumph-that-changed-the-world/ (Accessed on April 11, 2018)
26 *Computers in Translation and Linguistics*, A Report by the Automatic Language Processing Advisory Committee, Division of Behavioral Sciences, National Academy of Sciences, National Research Council, Publication No. 1416. www.mt-archive.info/ALPAC-1966.pdf (Accessed on October 26, 2017)
27 Marvin Minsky and Seymour Papert, 1972. "Artificial Intelligence progress report", *MIT Artificial Intelligence Memo* No. 252, Jan 1, 1972. https://dspace.mit.edu/bitstream/handle/1721.1/6087/AIM-252.pdf?sequence=2 (Accessed on October 26, 2017)

28 TD-Gammon is a computer backgammon program developed in 1992 by Gerald Tesauro at IBM's Thomas J. Watson Research Center. It explored strategies that humans had not pursued and led to advances in the theory of correct backgammon play. www.mitpressjournals.org/doi/abs/10.1162/neco.1994.6.2.215?journalCode=neco (Accessed on April 11, 2018)
29 Deep Blue versus Garry Kasparov was a pair of six-game chess matches between world chess champion Garry Kasparov and an IBM supercomputer called Deep Blue. The first match was played in Philadelphia in 1996 and won by Kasparov. The second was played in New York City in 1997 and won by Deep Blue. The 1997 match was the first defeat of a reigning world chess champion by a computer under tournament conditions. Deep Blue's win was seen as very symbolically significant, a sign that artificial intelligence was catching up to human intelligence, and could even defeat it. http://theconversation.com/twenty-years-on-from-deep-blue-vs-kasparov-how-a-chess-match-started-the-big-data-revolution-76882 (Accessed on April 11, 2018)
30 www.britannica.com/topic/World-Wide-Web (Accessed on April 11, 2018)
31 https://mitpress.mit.edu/books/perceptrons (Accessed on April 11, 2018)
32 www.technologyreview.com/s/603062/ai-winter-isnt-coming/ (Accessed on April 11, 2018)
33 Daniel Crevier, 1993. *AI: The Tumultuous Search for Artificial Intelligence*. New York, NY: Basic Books.
34 https://deepmind.com/research/alphago/ (Accessed on April 11, 2018)
35 www.ijcai.org/proceedings/2017/0772.pdf (Accessed on April 11, 2018)
36 M. Hilbert and P. Lopez, 2011. "The world's technological capacity to store, communicate and compute information". *Science*, 332(6025): 60–65. doi: 10.1126/science.1200970. www.martinhilbert.net/worldinfocapacity-html/(Accessed on October 26, 2017)
37 https://pages.experts-exchange.com/processing-power-compared (Accessed on October 26, 2017)
38 https://plato.stanford.edu/entries/connectionism/ (Accessed on April 11, 2018)
39 Frank Rosenblatt (1928–1971) was an American psychologist notable in the field of artificial intelligence. He conducted the early work on perceptrons, which culminated in the development and hardware construction of the Mark I Perceptron in 1960. This was essentially the first computer that could learn new skills by trial and error, using a type of neural network that simulates human thought processes. www.edubilla.com/inventor/frank-rosenblatt/ (Accessed on April 11, 2018)
40 https://appliedgo.net/perceptron/ (Accessed on April 11, 2018) http://computing.dcu.ie/~humphrys/Notes/Neural/single.neural.html (Accessed on April 11, 2018)
41 "Everyone is freaking out about artificial intelligence stealing jobs and leading to war – and totally missing the point." Kriti Sharma, Sage. Nov. 20, 2017. 43% of respondents to a Sage survey in the United States and 46% of respondents in the United Kingdom admitted that they have "no idea what AI is all about". Kriti Sharma is the Vice President of AI at Sage Group, a global integrated accounting, payroll and payment systems provider. She is also the creator of Pegg, the world's first AI assistant that manages everything from money to people, with users in 135 countries. www.businessinsider.in/Everyone-is-freaking-out-about-artificial-intelligence-stealing-jobs-and-leading-to-war-and-totally-missing-the-point/articleshow/61728469.cms (Accessed on February 14, 2018)
42 Krishna Reddy, "Advantages and disadvantages of Artificial Intelligence". https://content.wisestep.com/advantages-disadvantages-artificial-intelligence/ (Accessed on February 14, 2018). "Artificial Intelligence is designing machines that have the ability to think. It is the intelligence of machines. The discussions about the importance of artificial intelligence in our life have gained momentum in recent years. Is it a boon or a bane to future of human existence?, is an ongoing debate. The very idea to create an

artificial intelligence is to make the lives of humans easier. Researchers of artificial intelligence want to bring in the emotional quotient to the machines along with the general intelligence. Let us look into the pros and cons of artificial intelligence".

43 Francesca Rossi, "Artificial Intelligence: Potential benefits and ethical considerations". www.europarl.europa.eu/RegData/etudes/BRIE/2016/571380/IPOL_BRI%282016%29571380_EN.pdf (Accessed on February 14, 2018). "Like all powerful technologies, great care must be taken in AI's development and deployment. To reap the societal benefits of AI systems, we will first need to trust them and make sure that they follow the same ethical principles, moral values, professional codes, and social norms that we humans would follow in the same scenario. Research and educational efforts, as well as carefully designed regulations, must be put in place to achieve this goal".

2

WHY AI IS UNLIKE ANY OTHER TECHNOLOGY

> Deep Blue didn't win by being smarter than a human; it won by being millions of times faster than a human. Deep Blue had no intuition. An expert human player looks at a board position and immediately sees what areas of play are most likely to be fruitful or dangerous, whereas a computer has no innate sense of what is important and must explore many more options. Deep Blue also had no sense of the history of the game, and didn't know anything about its opponent. It played chess yet didn't understand chess; in the same way a calculator performs arithmetic but doesn't understand mathematics.
>
> —*Jeff Hawkins*[1]

The human brain is complex but we have begun to figure out how it works

If the human body is an evolutionary masterpiece, the human brain is a much less understood but a much bigger marvel. Not many would disagree with Isaac Asimov's characterization of the human brain as "the most complex and orderly arrangement of matter in the universe". The human brain with its intricate inner workings is an amazingly complex system, and is capable of doing much more than any software ever written by man. The amount of information that a brain can hold, process and pass on, and the speed with which this can be done, is unparalleled.

The body's in-built system computes and sends throughout the body billions of bits of information, information that controls every action, right down to the flicker of an eyelid. Inside the body, nerves are the wires that carry the information back and forth from the central nervous system. From ancient times, the human brain has fascinated scientists and philosophers alike, and early philosophers debated whether the seat of the soul was the heart or the brain. As per Aristotle,[2] the soul resided in the heart, and the brain's only function was to cool the blood.[3]

Much progress has been made since the times of Aristotle, but our understanding of the brain is still work-in-progress. There are many aspects of the brain, its functions and its dysfunctions, where our understanding is inadequate and incomplete. In addition, the research in the field has progressed in a nonlinear fashion. It is not unusual for new research findings to contradict previous widely held beliefs, old beliefs that slowly get replaced by the new findings. In the last several decades, researchers have been able to develop a good understanding of the physiology of the brain and there are models that explain how the circuitry of the brain helps move information around.

However, the progress in our understanding of serious degenerative ailments such as Alzheimer's disease, mental illness and depression has been frustratingly slow. But this does not mean that there are no reasons to feel optimistic about progress and how the field is moving forward. We have been successful in understanding several aspects about how the brain functions and there has been a significant amount of progress, especially in the last 60–70 years. We don't understand everything about the brain but we do understand much more than at any other point in human history.

Recent developments in brain research: Spectacularly groundbreaking

In a development which would have seemed like pure science fiction a few years ago, scientists have started studying human brain-to-brain communication. In a very simple but extremely interesting experiment in 2013, Chantel Prat, Andrea Stocco and other researchers of the Institute of Learning and Brain Studies, University of Washington, established brain-to-brain communication in humans.[4] They placed two people in front of a video game in two different parts of the college campus. The video game was started at the same time and progressed at the same speed for the two subjects.

However, the first person could watch the video game while the second sat with his back to it. The second person also wore noise-cancelling headphones and, therefore, had no idea about the progress of the game. Using an electroencephalography (EEG) cap, the brain signals of the first person were monitored and communicated across the campus to the second person. A transcranial magnetic stimulation coil was positioned above the second person's head, which could interpret the signals coming in from the first subject. The fascinating thing about this experiment was that the first subject's desire to shoot the gun in the video game could be communicated to the second subject who was not watching or listening to the game. The magnetic stimulation coil could convey the first person's thoughts to the second person and make his finger move to shoot the enemy in the video game.

In another development, researchers are finding significant links between gut bacteria and mental health.[5] A complex network of neurons filled with important neurotransmitters lines our intestines. This mass of neurons was previously mostly overlooked. But as scientists realize the importance of this set of nerves, technically called the enteric nervous system, some are even calling it the "second brain". While the main function of the enteric nerves is connected to digestion, scientists believe that it is far too complex to serve only this one limited function. There are increasingly solid and irrefutable signs that the enteric nervous system might have a significant role to play in the emotional well-being of a person.

Progress in the understanding of the brain has, therefore, been slow but steady. Discoveries in other fields such as physics, engineering, information technology, to name a few, have come together to help bring about new findings in brain studies. Indeed, the discipline has become more multidisciplinary than ever before. Research and development in AI will also interface with and gain from the studies on the brain in ways that we are only now beginning to appreciate.

Machines that (or who?) can think

Roughly speaking, AI is a machine that can display aspects of the brain's cognitive function. It might even improve and evolve into a more refined version of itself. This is a machine that can *think* independently without the need for human input or external prompting. The machine itself could be a computer, a robot, a virtual-reality based platform, or anything else that meets the basic definition that it can perform cognitive functions on its own.

As we discussed in Chapter 1, one possible way to create AI is to replicate aspects of the human brain, as in how the human brain thinks, how it reacts to situations, and how humans learn, solve problems, take decisions and react in emergencies. This means that AI is and will be different from all other previous technologies developed by humans. While all previous technologies were developed by humans to handle a specific task, the holy grail of AI is to artificially replicate and improve on the human brain.

It is possible that eventually AI machines will take over many tasks that currently require human intelligence and, hence, this technology will be different from anything else that humankind has ever developed. Some other characteristics that might make AI different are:[6]

- AI will have prescience, or the ability to predict. AI's prescience provides it with the ability to adapt its behavior according to unexpected events; no other technology is capable of this.
- AI is also capable of autonomy, which means that the system can take decisions about its future actions based on its internal states changing according to sensory data. This makes AI systems similar to biological systems.

- AI might self-improve. With machine learning, AI could become better every time it processes a new set of data.

It is very likely that when AI begins to learn on its own and self-improve without outside intervention, it will eventually maximize its autonomy. This could be positive and helpful under normal circumstances, but also raises an uncomfortable thought and possibility. If an AI system can continue to improve without any constraints and the system also has the autonomy to change and adapt, it might ultimately become independent. As the fictional story of Rossum's Robots reminds us, independent machines can be very helpful, but they might also become a hazard when they go awry.

AI: The building blocks

AI is driven by four basic characteristics. The first is complex algorithms that make an attempt to mimic the decision-making process based on multiple inputs and that have multiple optimization variables, including tangible and intangible factors. The second is the ability of algorithms to make use of huge amount of data, commonly known as BIG data,[7] data that is already available and that is being collected at an ever-increasing rate across the globe. The third is assessing the required computing capability that can use these algorithms and process this data.

These three are different parts of a complex jigsaw puzzle. What makes the picture complete is how these pieces fit together and is the most difficult and intelligent aspect, this is the fourth part of the entire equation. One needs to first define the problem at hand and then solve it in terms of the right mix of algorithms, assessing data and then efficiently process this data and, finally, estimate the computing power required and how that will be used to solve this puzzle.

Naturally, the "defining the problem" part is the most crucial. Unless one has defined the target "AI" problem well, it is extremely likely that it won't reach its logical conclusion. This is because the current forms of AI are weak, or in other words, specific to a defined purpose.[8] Current forms of AI focus on a specific problem, whether it is a "driverless car", "personal household assistant", "newspaper journalist", "column writer", "magazine correspondent", "shop floor worker", "customer care assistant", "restaurant worker", "music composer", "stock trader" or simply "someone you can talk to". As things stand today, a single machine or program will not be able to do multiple things without an upgrade or a modification. So at this stage, one AI machine can do a specific thing because it is designed to tackle a specific, target problem by using a method that is most appropriate for the task. For example, natural language processing or facial recognition technologies can be used in sentiment analysis, security applications, etc. There are AI solutions that are being built that target a problem, be it in diagnostics or market analysis and prediction. And in some cases, these solutions can be applied to multiple "problems" across sectors with some modifications and changes.

The types of possible AI

A machine can be called artificially intelligent when it is equipped with at least one human-like ability. One can be very general and say that anything, whether it is a machine or an algorithm or a robot, that can think and take decisions independently without the guidance of humans has Artificial Intelligence (AI), but there are many possible different types of specific AI:

Language based: This type of AI has the ability to speak and recognize languages. It has expertise in vocabulary and understanding of different accents in a language in order to decipher what is being said, something that could have otherwise been difficult. These machines can be used to interact as customer service agents and telephone callers. They can also be used to work as language teachers.
Mathematical ability: This specific AI has the ability to understand numbers and operations on numbers, with recognition of symbols used in mathematics and understanding of rules and conventions followed in the subject. These machines can be used to perform iterative, lengthy and complex calculations precisely. The more recent and complex machines can also be used in operations that require advanced functions including logic-based calculations.
Emotional intelligence: This is the ability to empathize with others and the ability to understand people's feelings. This is an area that is subject to debate in terms of how AI machines will handle these situations. There are machines that can respond to people's comments and statements but it is still doubtful whether the machines can read between the lines and understand the difference between what is being said and what is implied.
Reasoning and problem solving: This is the "reasoning" and "problem solving" ability, which is based on linking cause and effect. This is needed to make judgments and decisions on the basis of available information. The reasoning is needed to solve "optimization" problems when the problem involves multiple decision variables, the process to reach a solution and selecting the most suitable alternative. The specific AI machines can take "reasoning" decisions and solve the "optimization" problems as good as or even better than humans.
Self-improvement and learning: This is the foundation of civilization and the key to expanding knowledge frontiers for humanity. It is the ability to improve and learn from accumulated experience and knowledge to become proficient in an activity or developing a new skill. The ability of learning is unique to humans and there can be multiple sources of learning, including listening, reading and observation-based learning.

It is not just a coincidence that many of these areas are closely linked to the capabilities of the human brain. Because artificial intelligence is trying to mimic natural intelligence, it is expected that whenever scientists and researchers are working on specific AI, they are concentrating on a specific function of the human brain. The latest developments in AI have been helped by the advances we have made over the last few decades in understanding how the brain works.

AI: Interaction of many academic disciplines unlike other technology

Though the recent attempts to understand the brain may have belonged to the field of biology and medicine, substantial contributions have come from many academic branches. The tools and techniques developed in other areas and theoretical advances in other disciplines have helped in developing a better understanding of the brain and its functioning: so much so that it would be very difficult to claim today that brain studies entirely belong to any one particular academic discipline.

AI is a science focusing on the development of functions similar to human intelligence, such as cognition, reasoning, speech recognition, language skills, learning, problem solving and development of emergency responses such as "fight or flight". It is natural that as a science, it draws from many academic disciplines and calls for expertise in disciplines such as computer science, medicine, biology, psychology, linguistics, mathematics, and engineering.

AI is focusing on the development of computer functions that mirror human intelligence, such as cognition, compliance with socially acceptable practices, reasoning, persuasion, learning, knowledge transfer and problem solving. So, multiple areas can contribute to build an intelligent system: mathematics, biology, philosophy, computer science, sociology and neurology. This list is certainly not exhaustive and it is not unusual for other, less obvious, disciplines to also contribute towards AI development.

This is one of the important reasons why AI is unique and different to other technologies and why, at times, progress in AI is dependent on advancements in several disciplines.

AI is for decision-making; other technology is about better execution

Throughout human history, the focus of science has been on developing an understanding of natural phenomena and then identifying the laws of nature associated with them. For simple processes, this was done one by one and for complex ones, in conjunction with many others. Prehistoric man, through trial and error, figured out causal relations: that if X happens, Y will follow. These lessons were later on used to make predictive models for things around them. This predictive model could be used to develop useful technologies to make things that reduced human effort and improved the quality of life for humans.

Most of the focus of science and technology effort throughout human history has been on making the lives of human beings easier. The common thread in various technological endeavors has been that humans knew what they wanted and what they wanted to achieve. They were deciding the direction and were trying to figure out "how". They were taking the decisions on what is right and what is wrong, what is acceptable and what is not. Deciphering nature's laws and then putting limits on usage and practical applications was the path to progress.

In the entire process of scientific and technological achievements, humans have always been in control of "what" they want to achieve, "where" they want to reach and "how" they want to get there. The philosophy and final objective behind each and every scientific and technological invention has virtually remained the same: a safer, better, longer, happier and more comfortable life for humans. The objective has always been to reduce the misery of humankind that it is subject to either because of natural phenomena or because of human follies.

But AI could be different from all other scientific and technological developments so far. In the case of AI, humans are increasingly letting go of their control of decision-making. Humans are showing faith in machines and trust the non-human intelligence to make decisions on their behalf. Increasingly, these decisions are not only straightforward and simple ones but also more subjective ones, ones that require interpretation, ones that are ambiguous, and ones that do not have all the information and where the background is fuzzy.

In simpler terms, before AI, we were always in control of the thinking aspect; we never outsourced the "thinking" part to machines. However, this could change with AI. We are outsourcing the "thinking" part. Other technologies are similar to finding the most optimal path when we have already taken a decision about the direction, but AI is about both, the direction as well as the most optimal path. This is the key difference between other technologies and AI. ***In the context of AI, can thinking be outsourced, or more importantly, should thinking be outsourced, is the question facing all of us today.***

Other technologies handle predefined problems vs. AI handling unstructured challenges

Consider a large hospital where oncology patients are diagnosed by cancer specialists and put on the right course of treatment. Human experts usually make the final choice between invasive surgery and non-invasive methods for a particular patient. Human experts can use their experience and judgment when they see a case without precedence in their medical careers. Non-AI machines can only be helpful if a similar case exists in the database. But AI research is fast progressing in the direction where AI-based machines will be able to take decisions just like any other human expert based on data and experience.

Most technologies have a specific purpose and are given a predefined problem that needs to be solved. The task or the problem could be as mundane as doing tricky mathematical calculations to as complex as flying an aircraft and making weather forecasts. However, irrespective of the situation, technological advancements have a predefined and well-articulated purpose, function and cost.

A machine, a simulation or a computer program can answer the specific questions and handle the problems it is meant to solve. In most cases in history, scientific discovery and innovation has followed a linear path where the solution may have been invisible but the problem was clearly visible and definable. Even if the path was not known, the starting point and destination were usually visible to researchers.

The case of AI is different. Apart from solving simple or complex well-defined problems and doing specific tasks, machines with AI can also answer unstructured and generic questions in their area of expertise. The objective and response mechanism for an AI machine doesn't need to always be predefined and could even fall outside the realm of possibilities imagined at the time of its design. Non-AI programs and machines are only able to handle the situations that they have been programmed for and that they are familiar with. However, in principle, AI-based machines would not have this constraint.

As an example, there are many ancient scripts that experts have not been able to infer and understand.[9] Most types of tools and technologies may only have limited usage in helping to decipher these scripts. However, AI can help solve such puzzles much more effectively and may prove helpful in interpreting the pattern and symbols of a language that we don't understand today. AI, in a manner, is similar to many other human experts working on languages. While a non-AI machine and translator can help make tasks simpler, AI machines can go way beyond this.

AI-based industries: Winner takes all

How many auto companies can you name? Of course, there are many including Toyota, General Motors, Daimler, Hyundai, Honda, Nissan, Ford, Suzuki, Volkswagen and BMW. How many consumer companies can you name? Again, there are many of these including Unilever, Procter & Gamble, Reckitt Benckiser, Nestle, Johnson & Johnson, Coca Cola, Pepsi, Philip Morris and Mondelez. How many pharmaceutical companies can you name? Once again, there are many of these and they include Pfizer, Astra Zeneca, Glaxo SmithKline, Sanofi, Merck, Bayer and Roche.

Now, let us see some other industries.

How many competitors of Google can you name? In a way, there are many including eBay, Yahoo, AOL, Expedia, Facebook, and in other ways, there are none. Even Eric Schmidt has said that Amazon or Apple are the closest competitors to Google.[10] If we look at only the search engine business, Baidu from China is the closest competitor but that is more because the Chinese government makes it difficult for companies in the information dissemination business to operate in China. However, no other company has the scale of Google.

How many competitors of Facebook can you name? These would be Instagram (which is wholly owned by Facebook), Snapchat and LinkedIn in various ways. There is also WeChat in China and Vkontakte in Russia. But again, none of these have the scale of Facebook.

How many competitors of WhatsApp can you name? How many competitors of Amazon can you name? How many competitors of Alibaba can you name? How many competitors of Microsoft can you name? How many competitors of LinkedIn can you name? How many competitors of Instagram can you name? How many competitors of YouTube can you name?

One thing is very clear. The new economy companies neither belong to a people-intensive industry nor are the new businesses conducive to competition. The case of AI is not very different. Just like other internet-based businesses, AI is a "winner takes all" industry. There is absolutely no space for smaller competitors. Small and large competitors cannot survive in "peaceful coexistence". In an internet-based business, a company that is two times larger (not because it is better or has a better product or service offering) than its nearest competitor will grow three or four times larger in a short span of time. Sooner or later, the larger competitor will elbow out the smaller competitor.

We can have many large Fast Moving Consumer Goods (FMCG) or auto or pharmaceutical companies but it is doubtful if we can have many large Amazons or Facebooks. There can be many successful multinational companies in old economy industries. And similarly, at a local level, every country can have many local FMCG, automobiles and pharmaceutical companies that are influential and can do good business. So, these industries from the old economy are decentralized. However, only a handful of elite companies would be able to keep majority gains in AI-based industries and power will be concentrated only in the hands of a few corporations.

Can AI be creative?

We have discussed three characteristics of AI: self-improvement, prescience and autonomy. A natural question then is to ask if AI can be creative. Perhaps, more than anything else, creative expression is one activity that uniquely differentiates humans from other living creatures. Humans can create art, intuit mathematical expressions, and occupy various realms that are part art and part science. For example, Vincent Van Gogh and Srinivas Ramanujan, Kurt Gödel and M C Escher, William Shakespeare and Steven Spielberg, Stephen Hawking and Akira Kurosawa, Albert Einstein and Richard Feynman; they all explored and expressed creativity in several different ways.

Creativity is not easy to mathematically define, nor is it easy to predict who the next Pablo Picasso will be. Creativity seems to safely reside in the bastion of human beings. And the critics of AI have pointed towards creativity as an unachievable goal for machines, irrespective of how intelligent they are. The critics of AI and skeptics of the lot claim that a Deep Blue can defeat Gary Kasparov or IBM Watson can win Jeopardy,[11] but it is doubtful if an AI machine can postulate the General Theory of Relativity like Einstein or can author *War and Peace* like Leo Tolstoy. While AI is nowhere close to producing the next AI-Picasso, it is instructive to explore the small but firm progress being made in this direction by AI machines.

The artists

Art has always been considered a universal expression of creativity. From the earliest known art work of the Old Stone Age, to Picasso and Andy Warhol, art and creativity have always been synonymous with each other. Who is to say whether the Stone Age artists were more or less creative than Van Gogh or Rembrandt? Creativity is not confined to a particular region or particular era. Throughout history, art competitions have served the important purpose of inspiring and supporting budding artists. Contests have helped find undiscovered talent.

One such competition took place in 2016[12] and has become an annual feature now. Artists competed from all over the world. Some of the goals of the competition were to (a) encourage participation by the public, (b) challenge students to apply skills in creative ways, and (c) integrate aesthetics and technology. There were also stringent rules to be followed by the competitors. There were two categories: an Original Artwork category for the type of art where no specific inspirational reference image or material is used, and a Re-interpreted Artwork category for the type of artwork painted using a reference image such as a photo or an image of a famous painting. Each team could submit up to six distinct artworks in each of the two categories. Rules pertained to the coloring material as well. Paint or color had to be applied with physical brushes. Art made by inkjet-like matrix printer would have led to disqualification.

The competition was intense. When the results were declared in May 2017, the winners were from various parts of the world, proving again that creativity is not confined to a particular place. The first prize, $30,000, went to Taida from Taiwan. Taida had the style of a classical painter. It mixed its own palette from a limited selection of color paints. After achieving the color it desired, Taida worked in layers to achieve the effect it wanted. The second prize of $18,000 went to a painter from the United States who took several photos, selected its favorite photo, and made a portrait based on the photo one paint stroke at a time. The third prize of $12,000 went to NoRAA from Italy. While the other artists focused on representational artwork, NoRAA's art was abstract and very popular with the professional judges.

This art competition might have been similar to many others with the goal to encourage art and creativity among unrecognized artists. However, what made this competition especially interesting was that all the competitors were robots. The three winners were AI. While debate will continue regarding the nature and origin of creativity, it is worth noting that creativity itself can defy definition. Creativity, and the definition of creativity, can differ from one field to another. Music, art, mathematics, sciences all involve creativity in its multiple facets. Who is to say that AI is incapable of creativity and an appreciation for beauty?

Deep dreaming

We have all daydreamed. Children are especially adept at it. Looking at a picture in front of them, when asked what they are seeing, it is likely that they will describe seeing patterns or objects that adults do not see. Children's minds are creative and imaginative. Can AI also similarly daydream? The answer, surprisingly, seems to be yes. A machine learning program developed by Google, aptly called DeepDream,[13] can be shown a picture and asked what it is seeing. And the program then produces what it actually sees. What it produces has been described as a sort of post-impressionistic art inspired by the original picture. DeepDream seems to be the AI equivalent of a budding Van Gogh that looks at a landscape and comes up with its own vision. A vision that seems quite unlike what most people see, but that seems to contain creativity in each stroke of the brush. AI, the artist type, seems to have a mind of its own!

Mind meld

The possibilities that can unfold as AI and humans interact can really push the bounds of imagination. AI will impact and interact with humans unlike any other technology. The only danger is for the majority of the population not to take the possibilities seriously. Reality can catch up with fiction faster than we think. Let's begin with fantasy. Superhero Tony Stark, from the *Iron Man* movies, owes all his powers to the arc reactor, the powerful electric generator. Tony is a genius scientist, but wouldn't be able to achieve his heroic feats without the help of J.A.R.V.I.S. (Just Another Rather Very Intelligent System). Iron Man is not just Tony Stark or just J.A.R.V.I.S.; rather, he is the result of the collaboration between Stark's human intelligence and J.A.R.V.I.S.'s AI.

Now let us come to what could soon be reality. Elon Musk, who was inspiration for the *Iron Man* movies, has a new vision for humans and AI.[14] According to this vision, the only way to prevent a final showdown between humans and AI is to create symbiosis between the two. Just like it exists between Tony Stark and J.A.R.V.I.S. and perhaps even more than that. In order to create this symbiosis, as per Musk, one needs to go beyond verbal communication between humans and AI. Humans can emit around 40 bits per second while speaking; Musk considers this too slow. Not good enough, although it works for Iron Man. Musk established a start-up company, called Neuralink, in 2017 to solve this problem and speed up communication between man and machine. The company plans to develop methods to speed up this communication via brain implants. The idea is for humans to not speak, but communicate to AI via these implants. This is not to say that this won't turn out to be a very difficult, if not impossible, problem to solve. But, Musk is not alone in pursuing such ideas. Sam Altman, the president of Y-Combinator, is also a believer in the possibility of a future where humans and AI merge in some form or another. AI is forcing us to think of future potentialities that sometimes make fiction appear tame.

Notes

1 Jeffrey Hawkins (born in 1957) is the American founder of Palm Computing (where he invented the PalmPilot) and Handspring (where he invented the Treo). He has since turned to work on neuroscience full time, founded the Redwood Center for Theoretical Neuroscience (formerly the Redwood Neuroscience Institute) in 2002, founded Numenta in 2005 and published *On Intelligence* describing his memory-prediction framework theory of the brain. https://redwood.berkeley.edu/ (Accessed on April 11, 2018)

2 Aristotle, Greek Aristoteles (born 384 BCE, Stagira, Chalcidice, Greece – died 322 BCE, Chalcis, Euboea), ancient Greek philosopher and scientist, one of the greatest intellectual figures of Western history. He was the author of a philosophical and scientific system that became the framework and vehicle for both Christian Scholasticism and medieval Islamic philosophy. Even after the intellectual revolutions of the Renaissance, the Reformation, and the Enlightenment, Aristotelian concepts remained embedded in Western thinking. www.britannica.com/biography/Aristotle (Accessed on November 21, 2017)

3 Charles G. Gross, 1995. "Aristotle on the brain". *The Neuroscientist*, 1(4). www.princeton.edu/~cggross/Neuroscientist_95-1.pdf (Accessed on November 20, 2017)

4 Rajesh P. N. Rao, Andrea Stocco, Matthew Bryan, Devapratim Sarma, Tiffany M. Youngquist, Joseph Wu, and Chantel S. Prat, 2014. "A direct brain-to-brain interface in humans." *PloS one*, 9(11): e111332; www.nbcnews.com/science/mind-meld-scientist-uses-his-brain-control-another-guys-finger-8C11015078(Accessed on November 20, 2017)

5 Jane A. Foster and Karen-Anne McVey Neufeld, 2013. "Gut–brain axis: How the microbiome influences anxiety and depression". *Trends in Neurosciences*, 36(5): 305–12; David Kohn, "When gut bacteria change brain function", *The Atlantic*, Jun. 24, 2015. www.theatlantic.com/health/archive/2015/06/gut-bacteria-on-the-brain/395918/ (Accessed on November 20, 2017); www.sciencedaily.com/releases/2017/08/170821122736.htm(Accessed on November 20, 2017)

6 George Zarkadakis is a contributor at HuffPost and he is an AI engineer and writer. See his article on the website of the *Huffington Post*: "The 3 things that make AI unlike any other technology". www.huffingtonpost.com/entry/the-3-things-that-make-ai-unlike-any-other-technology_us_58f25764e4b0156697224fcc (Accessed on October 24, 2017)

7 "Big data" is a term for data sets that are so large or complex that traditional data processing application software is inadequate to deal with them. Big data challenges

include capturing data, data storage, data analysis, search, sharing, transfer, visualization, querying, updating and information privacy. Lately, the term "Big data" tends to refer to the use of predictive analytics, user behavior analytics, or certain other advanced data analytics methods that extract value from data. www.sas.com/en_in/insights/big-data/what-is-big-data.html (Accessed on April 11, 2018)

8 Weak artificial intelligence (weak AI), also known as narrow AI, is artificial intelligence that is focused on one narrow task. Weak AI is defined in contrast to either strong AI (a machine with consciousness, sentience and mind) or artificial general intelligence (a machine with the ability to apply intelligence to any problem, rather than just one specific problem). All currently existing systems considered artificial intelligence are weak AI. https://searchenterpriseai.techtarget.com/definition/narrow-AI-weak-AI (Accessed on October 24, 2017)

9 http://mentalfloss.com/article/12884/8-ancient-writing-systems-havent-been-deciphered-yet (Accessed on October 25, 2017)

10 Amazon may not be a direct competitor to Google. In Berlin in 2014, Google's Executive Chairman, Eric Schmidt, had this to say: "Many people think our main competition is Bing or Yahoo. But really, our biggest search competitor is Amazon." The main area of competition for these two companies is product searches. Eric Schmidt has cited other competition along the way. For example, in an interview with Bloomberg, he indicated that the competition with Apple is the "defining fight of the computer industry." https://searchengineland.com/competitive-threats-google-means-249772(Accessed on October 25, 2017)

11 An interesting example of cognitive computing is IBM's Watson, which is far, far ahead in "subjective thinking" and a multiple-times improved version of the machine (largely a machine that was only good in "objective or logical thinking") that beat Gary Kasparov in chess. From Martin Ford, 2015. *The Rise of the Robots: Technology and the Threat of Mass Unemployment*. London: OneWorld/Pan Macmillan, p. 334. www.ibm.com/watson/ (Accessed on April 11, 2018)

12 https://thenextweb.com/insider/2016/04/22/worlds-next-great-artist-might-robot/, The RobotArt is a $100,000 Robot Art competition now. https://robotart.org/ (Accessed on November 20, 2017)

13 DeepDream is a computer vision program created by Google engineer Alexander Mordvintsev that uses a convolutional neural network to find and enhance patterns in images, thus creating a dream-like hallucinogenic appearance in over-processed images. www.telegraph.co.uk/technology/google/11730050/deep-dream-best-images.html (Accessed on April 11, 2018)

14 Neuralink is an American neurotechnology company founded by Elon Musk and eight others, reported to be developing implantable brain–computer interfaces (BCIs). The company's headquarters are in San Francisco; it was started in 2016 and was first publicly reported in March 2017. The trademark "Neuralink" was purchased from its previous owners in January 2017. www.neuralink.com/ (Accessed on April 11, 2018); Dana Hull, "Elon Musk's Neuralink gets $27 million to build brain computers", *Bloomberg Technology*, Aug. 25 2017. www.bloomberg.com/news/articles/2017-08-25/elon-musk-s-neuralink-gets-27-million-to-build-brain-computers (Accessed on November 20, 2017)

3

EMERGING DANGER OF AI-INDUCED MASS UNEMPLOYMENT

The automation of factories has already decimated jobs in traditional manufacturing, and the rise of artificial intelligence is likely to extend this job destruction deep into the middle classes, with only the most caring, creative or supervisory roles remaining.

—*Stephen Hawking*[1]

What to do about mass unemployment? This is going to be a massive social challenge. There will be fewer and fewer jobs that a robot cannot do better [than a human]. These are not things that I wish will happen. These are simply things that I think probably will happen.

—*Elon Musk*[2]

You cross the threshold of job-replacement of certain activities all sort of at once. So, you know, warehouse work, driving, room cleanup, there's quite a few things that are meaningful job categories that, certainly in the next 20 years will go away.

—*Bill Gates*[3]

There are lots of examples of routine, middle-skilled jobs that involve relatively structured tasks, and those are the jobs that are being eliminated the fastest. Those kinds of jobs are easier for our friends in the artificial intelligence community to design robots to handle them. They could be software robots; they could be physical robots.

—*Erik Brynjolfsson*[4]

Automation is already a giant job killer

In her December 2016 *New York Times* article "Long-Term Jobs Killer Is Not China. It's Automation", Claire Cain Miller highlighted the impact of automation on jobs.[5] She wrote:

> Take the steel industry. It lost 400,000 people, 75 percent of its work force, between 1962 and 2005. But its shipments did not decline, according to a study published in the American Economic Review last year. The reason was a new technology called the minimill. Its effect remained strong even after controlling for management practices; job losses in the Midwest; international trade; and unionization rates, found the authors of the study, Allan Collard-Wexler of Duke and Jan De Loecker of Princeton.
>
> Another analysis, from Ball State University, attributed roughly 13 percent of manufacturing job losses to trade and the rest to enhanced productivity because of automation. Apparel making was hit hardest by trade, it said, and computer and electronics manufacturing was hit hardest by technological advances.

The impact of automation has been selective so far, particularly impacting certain types of jobs. For example, jobs and wages for workers without college degrees doing manual labor have not really recovered yet. But, increasingly, the changes have become more widespread, affecting white-collar and service-sector workers too.

Automation is already a big threat to the transportation and warehousing industry, which employs 5 million Americans; 2.5 million drivers and 8 million Americans work in the retail salespeople industry. Many of these jobs already are or will be automatable in the not-so-distant future. Another industry under risk is QSRs (Quick Service Restaurants); restaurants in the US employ 14 million workers but they operate on wafer-thin margins, and many functions such as taking orders, serving food and washing dishes are at risk because of automation.

Smarter machines that can do more

While automation and robots put low-skill jobs at risk, there are many high-skill jobs as well that are at risk because of AI. It would be a mistake to think that computers can only do what they are programmed for. They are entering the erstwhile exclusive "humans-only" domains, which have subjective elements. Some of the tasks that were simply unthinkable for AI a few years back can be done by AI almost as well as or better than humans can do.

For example, newspaper and magazine articles can be written by advanced algorithms. AI can also suggest the best possible medical treatment and is no longer limited to scanning patient records. Another interesting example of cognitive computing is IBM's Watson, which is as adept at "subjective thinking" as it is at "objective thinking".[6]

IBM has developed an advanced cognitive system that is capable of answering questions posed in natural language. This is called Watson and was developed in IBM's DeepQA project by a research team led by principal investigator David Ferrucci. Watson was named after IBM's first CEO, the industrialist Thomas J. Watson. The computer system was specifically developed to answer questions on the quiz show *Jeopardy*. In 2011, Watson competed on Jeopardy against former winners Brad Rutter and Ken Jennings, and defeated them.

Watson had access to 200 million pages of structured and unstructured content consuming a gigantic 4 terabytes of disk storage. In February 2013, IBM announced that the Watson software system's first commercial application would be in lung cancer treatment at Memorial Sloan Kettering Cancer Center, New York, in conjunction with the health insurance company WellPoint. On IBM's homepage on Watson, the company claims that Watson APIs and solutions are currently in use in 45 countries.

What does history tell us?

Throughout modern human history, and especially since the Industrial Revolution, various experts have warned about the impact of machines on employment opportunities. But, before the beginning of the 21st century, these fears were always baseless. More technology always meant more jobs, better-paying jobs, and as a result, a more prosperous society. Yes, it is true that change in technology may have led to a short-term unemployment crisis during the transition phase in a few segments, but this never became a structural challenge for long. The rise in wealth and available opportunities were always moving in one direction apart from momentary blips.

But, as history has taught us, "something like this has never happened ever" is not a guarantee that it will not happen in the future. For example, the almost equal relationship of 30 years between higher productivity and higher wages in the post Second World War (WWII) period ended around the middle of the 1970s for developed countries. For more than the last 40 years, the average worker has been earning much less in real terms despite more than double the increase in productivity.

In the United States, the first decade of the 21st century saw absolutely no new jobs. Even during the late 1920s and early 1930s, the situation was better. This time around, technology is not only threatening "routine" and algorithmically definable jobs but also impacting jobs that were considered safe earlier. From the worker's perspective, the unfortunate part is that more education and skills are unlikely to help much and even some specialist functions can be performed better by machines than humans.

Another big challenge is sub-optimal employment. AI could also lead to not just "unemployment" but chronic "underemployment" for over-qualified workers through no apparent fault of their own; e.g. many university graduates have to take jobs that may not require a degree. The irony of higher education is that the more useless it gets; the more and more you will need it.

How will AI impact jobs?

There is still no consensus as to how AI will impact the number of available jobs in an economy. Nevertheless, there is a general agreement that AI is a disruptive technology. Even some of the people who believe that AI will not be the reason for a drastic cut in jobs that only humans can perform today agree that AI will lead to a shift in the nature of jobs; the implication is that there will also be a transformation in the relative importance of skills. It can be safely assumed that as the adoption of AI increases in the economy, we will need to change our skilling and training initiatives. To make ourselves relevant for jobs in an AI era, there will be a need for people to adapt and collaborate with AI. The most common understanding is that the first target of AI will be jobs that are monotonous, or in other words, those that can be called programmable, but it is unlikely to remain limited to that. So, low-skill jobs or high-skill jobs, boring jobs or creative jobs, the need for re-skilling the workforce will be universal because of the changes AI will bring.

- Depending on which argument sounds more convincing to you, there are six possibilities:
- AI will be responsible for a significant number of job cuts (between 25% and 50%) in the next 20 years. The cost savings will largely remain with big corporations and ordinary people will be worse off. *This possibility looks politically unviable.*
- AI will be responsible for a significant number of job cuts (between 25% and 50%) in the next 20 years. But, the national governments individually and collectively will find a solution to ensure that this cost saving comes back to society. *This requires political intervention; a similar sentiment is reflected in the election of Donald Trump in the US and the UK's Brexit vote.*
- AI will impact a much lower number of jobs (less than 10%) but the impact will be much stronger in those jobs that are low-skill jobs. This implies that people from lower economic strata will be affected more. *This will lead to higher inequality in society.*
- AI will impact a much lower number of jobs (less than 10%) but the impact will be across the board, implying that low/high-skill jobs are equally at risk because the threshold for investment required to replace humans is much higher in high-skill jobs. *This is less bad than the first three scenarios.*
- The most optimistic scenario is that AI will lead to more jobs. In this scenario, AI will not really be different than the previous waves of

industrialization and technology shift. *The initial signs are not really encouraging that this will be the case.*

- In the last 50–60 years, AI has consistently failed to live up to its promise and an excessive interest in AI has been followed by a long AI winter (a phase when AI loses favor among scientists and researchers, media and corporations). It won't be different this time as well. *The enabling environment has changed for AI and it is unlikely that the future will be the same as the past.*

All of these scenarios rest on strong arguments, but it would be quite premature to brush AI off and believe that it is business as usual. AI may not be able to take away jobs entirely but it can certainly reduce the effort involved, which means that there will be less people required for the same quantum of work. The potential of AI to make a job more efficient for humans is immense.

Businesses will become more efficient and leaner. Hence, the demand for manual work may fall, especially on the shop floor and at the lower levels. But, highly skilled professionals like doctors, lawyers, journalists and IT specialists are also likely to see an impact of AI on jobs.

It is possible that AI will augment the work that humans perform. But, it would be foolish to ignore its disruptive potential, especially in areas where job descriptions could be fully automated and AI could replace humans. No one really knows at this stage how AI will impact jobs, but hopefully the active discussion on this will help in moving the debate forward and in finding solutions. However, people have already started to think about how humanity and governments should respond to "AI's threat to jobs". An interesting solution that is being advocated by several experts is offering a basic income to sustain survival. It is most commonly called "Universal Basic Income" and will be paid to people irrespective of their job status, i.e. whether they are in employment or unemployed. In some countries, this discussion has reached an advanced stage; for example, though the idea was finally rejected in a vote, Switzerland offered the option of having Universal Basic Income to its people. Even in developing and poor countries such as India, some of the senior government officials have lobbied for Universal Basic Income to be considered by the government, at least for the poorest of the poor. This is not linked to AI specifically, but the core philosophy is the same.

The impact is not just limited to the developed world

It is not just the developed world that is adopting AI-based applications to save costs and eliminate the need for human intervention; countries like China and India are also witnessing this change. For example, the Banking, Financial Services and Insurance (BFSI) sector is adopting AI-powered chatbots for use as virtual agents. Healthcare is another sector where there are a number of AI solutions being developed in India and a number of the large IT companies in India have come up with their own AI platforms.

Let us discuss the banking sector in a little more detail because this is one area that has implications for almost everyone on a daily basis. India has a large-scale presence of the national government in the banking system, and several banks are majority owned by the government, including the State Bank of India (SBI), which is the largest bank in India. Technology adoption is in general slower in the state-owned companies but some of the private-sector companies are increasingly relying on AI-based technologies and see it as the future.

For example, HDFC Bank, which is one of the largest banks in India and the country's leading and more than 20-year-old private-sector banking institution, is at the forefront of leading this initiative. HDFC Bank said in September 2017 that its Facebook Messenger chatbot (OnChat) had registered a 160 percent month-on-month growth in transactions. HDFC Bank also has an AI chatbot called "Eva", which has addressed over 2.7 million customer queries in the six months since its inception. Others are not far behind. The banking app of DBS Bank has in-built AI. ICICI Bank, IndusInd Bank, Tata Capital and Yes Bank are also increasingly adopting more and more AI-based technologies to handle not only back-end operations work but also customer-facing functions.

The state-owned enterprises are trying to catch up. In September 2017, the State Bank of India (SBI) launched an AI-driven chat assistant called SBI Intelligent Assistant (SIA) to address customer queries. SIA is extremely powerful and is hardly going to have any capacity constraints in the near future as it can handle approximately 10,000 enquiries per second. The thinking behind an app like SIA is that most of the customer queries are routine in nature and can be handled by a chatbot. Typical queries involve product information on retail offerings like home, education, four-wheeler, two-wheeler and personal loans (asset side for banks) and information on the liabilities' side (savings and term deposits).

One of the authors of this book (Pankaj) has a banking relationship with HDFC Bank and he has noticed that the frequency of mailers on product offerings and updates on new promotions and partnerships, which are extremely localized and personalized, has increased considerably over the previous 12 to 18 months. With this, customer engagement has not only increased from the bank's perspective but it has also become much more useful from the customer's perspective. Most of these initiatives are based on automated technologies and this is clearly an indication that a bank's customer interface will be more and more dependent on them.[7]

While a lot of recent debate on the jobs impact of AI has focused on the developed world, it has become increasingly clear that AI and automation will have a tremendous impact on the developing world as well. The human impact is arguably much greater, due to the sheer number of people that AI can impact and disrupt in developing economies. We consider two developing world cases where the effect of automation is already being felt: the Indian IT industry and the Chinese manufacturing industry.

The case of the Indian Information Technology industry

26 is an interesting number; it is the only integer that is one greater than a square ($5^2 + 1$) and one less than a cube ($3^3 - 1$). "26" is also relevant for demographics; 26 years is the approximate "median age" of the Indian population. This means that half the people (almost 650 million in absolute terms) are below this age and there are 400 million people in the age group 10–24 years. The total number of people in the age group 45–64 years is approximately 150 million.

India needs 12–15 million new jobs every year

As per a joint report by the Confederation of Indian Industry (CII) and the Boston Consulting Group (BCG) titled *India: Growth and Jobs in the New Globalization* released in March 2017, over the next five years India will need to create 12–15 million non-agricultural jobs per year. However, between 2005 and 2012, only 8 million such jobs were created. Hence, the gap of 4–7 million jobs a year needs to be addressed, which is likely to rise with the increase in the number of young people joining the labor force.[8]

When we look at the percentage distribution of the population across age groups and more specifically in the 0–14 age group that accounts for 30 percent of India's total population, it is clear that India needs several million jobs over next few decades.[9] As education continues to reach more and more people and the number of people with technical and professional qualification continues to increase, India will also need a lot more white-collar jobs. One can question the percentage participation of women in the workforce or how many people are actively seeking employment, but the fact remains that the need for job creation in the Indian economy is huge.

These estimates could vary depending on the source but one thing is absolutely clear: the Indian economy has not been able to provide enough good-quality jobs to people entering the workforce. For example, the Asia-Pacific Human Development Report 2016 by the United Nations Development Program (UNDP) titled *Shaping the Future: How Changing Demographics Can Power Human Development* says that by 2050, more than 280 million more people will enter the job market in India alone, yet between 1991 and 2013, the economy absorbed less than half of new entrants to the labor market.[10]

How serious the job problem is can be understood from the fact that more than 25 million people applied for about 90,000 positions in Indian Railways, when one of the largest public-sector employers in India decided to hire a significant number of employees to expand its workforce to make its operation more efficient. This means that for each job, there were more than 250 applicants, and this is when these jobs were not extremely well paying and mostly belonged to the lower categories.[11]

Hiring by Indian IT is in trouble

In the context of the need for job creation in the Indian economy, there are a few messages emerging from the Indian IT industry:

1. Indian IT companies are hiring fewer employees. India's five largest companies—Tata Consultancy Services Ltd (TCS), Infosys Ltd, Wipro Ltd, HCL Technologies Ltd and Tech Mahindra Ltd, and Nasdaq-listed Cognizant Technology Solutions Corp.—together employed 1,243,777 people at the end of September 2017, down from 1,247,934 people at the end of March 2017, translating to a net reduction of 4,157 people. These six companies, which together employed more than a quarter of the 3.9 million people working in the industry, added 59,940 people in the first half of 2016–17, i.e. during April to September 2016.[12]
2. Automation is already impacting entry-level jobs. Industry leaders such as the ex-CEO of Infosys, Vishal Sikka, believe that India will see the biggest negative impact on jobs due to automation.[13]
3. The IT companies need to increase hiring in the geographies they operate in because of concerns over visa issues. The most prominent among them is the United States.[14]

The IT industry matters for the hiring of white-collar workers

India produces about 1.6 million engineers every year,[15] and for many years the IT industry absorbed about 200,000 new employees. This is under serious threat now. We will talk about India's Gini coefficient and rising inequality some other time and the comparisons may look premature at this juncture, but more often than not in history, rising inequality and its violent repercussions have their roots in lack of opportunities. For example, the often-cited cause of the French Revolution of 1789 was the massive disparity created among different classes of population due to political mismanagement, expensive wars and prolonged economic difficulty for the masses.

A huge number of educated yet unemployed youth with high aspirations is not the right ingredient for the India Growth story. This also puts a serious question mark on India enjoying a huge demographic dividend in the next couple of decades. It could also be the primary reason why job creation in "manufacturing" will be as important as job creation in "services" for the Indian economy in the next few decades,

In the Indian IT industry, the balance has decisively shifted in favor of employers over the last several years, and the trend in wages has been negative regarding starting salaries and hikes for junior employees. Indian IT industry veterans such as T V Mohandas Pai and Infosys co-founder N R Narayana Murthy believe that the salaries of freshers in the software industry have stayed stagnant, while those of senior-level employees have grown multi-fold.[16]

The Indian IT industry is staring at the prospect of a much lower growth figure, which will be called par, and the situation is very likely to get worse. There is already enough (and continuously increasing) supply available and demand is diminishing (as seen in the commentary from most IT companies), which may be good news for you if you are hiring but is a huge challenge to overcome if you are a knowledge worker.

Why education will cease to be the ticket for a better life

Consider the following: a) the situation is so bad that AICTE is cutting down the number of engineering colleges and the intake of students;[17] b) approximately 80 percent of India's engineers and more than 90 percent of MBA graduates are not employable and need extensive training to become job-ready,[18] c) the starting salaries of fresh graduates (if you move away from Tier I colleges) are either stagnant or have grown at a significantly lower rate than inflation,[19] and d) people in the top tier are seeing their compensation growing at a brisk pace and the gap between the top-end and the bottom is getting wider.

What about affordability? If we look at most professional courses in India, the cost of education has increased significantly.[20] The situation in India may look a little too bad but even in developed countries where the supply is more controlled, the opportunities available for people graduating with university degrees have been growing at a much slower rate. So, there are a lot of university graduates with professional qualifications who are not being optimally employed despite doing everything correctly. The situation is not going to improve soon and AI could even make it worse.

Hence, it seems very likely that at least for a few decades to come, a) more people with impeccable academic credentials will be available in the job market; b) higher education will continue to be unaffordable and the debt burden on freshly minted graduates will go up; c) opportunities will not keep pace with supply; and d) employers will have more options and at a cheaper cost to fill entry-level positions. Return on Investment (RoI) on higher education will continue to diminish and will be more unpredictable and inconsistent for an individual. However, skipping higher education would be much worse. Since almost everyone is getting more and more qualified, the "qualifying criterion" of a higher education degree to be even considered for a suitable opening will get become more stringent. The view will be fantastic from the top but in percentage terms, less people will reach the summit.

The implications are that: a) companies, especially in the service industry, have much less pressure regarding employee costs; b) the attrition could remain high at the bottom of the pyramid since wages will not be much different vs. other options available to employees; c) employee productivity (revenue vs. cost per employee) could increase; d) profitability as a percentage of the top-line may improve if everything else remains the same; and e) specialized high-end manufacturing will see renewed interest from prospective employees.

So, it will continue to be a great time for anyone who has "employee costs" as a huge contributor to the total spend, e.g. IT and BFSI companies will benefit and as a result, they will see better margins. Manufacturing companies may have less trouble in attracting talent but since the employee cost is much less as a proportion, the profitability impact would be limited.

Foxbots of Foxconn

Foxconn Technology Group is a Taiwanese multinational that carries out contract manufacturing projects for various electronics and IT companies. It has an impressive list of clients such as iPhone, iPad, Kindle, Nintendo, PlayStation, and a number of other well-known companies. Foxconn is the world's largest contract electronics manufacturer and the fourth largest IT company by revenue. It employs over a million workers in its operations, many of whom work in giant almost city-like facilities that are known locally by names such as iPhone City. Foxconn's iPhone City, located in China's Zhengzhou region, can produce up to 500,000 iPhones a day, an impressive number. Locals nowadays refer to Zhengzhou as simply "iPhone City".[21] Foxconn also happens to be China's largest private employer.

China is an attractive destination for companies, where they can not only lower their manufacturing costs, but also access the huge consumer market. Not unlike other countries, the local officials in China came up with incentives for the companies to make it easier and cheaper to do business. For decades, Beijing has encouraged such efforts by developing special economic zones that offer tax breaks to multinationals and relieve them from unnecessary rules.

Foxconn's manufacturing operations and facilities require a well-trained workforce. In the case of Zhengzhou, the local government plays a big role in sourcing a steady stream of workers. As a part of the deal with Foxconn, it recruits, trains and houses the workers. The local officials actively conduct recruitment drives for workers in nearby towns and villages. This provides for a good symbiotic relationship between Foxconn and the local governments and takes care of the employment of millions of workers. It is, therefore, not a surprise that China provides huge incentives to Foxconn.

2010 labor unrest

The year 2010 saw a number of labor protests and strikes in the southern part of China in response to the poor work conditions and low wages. These protests started at Honda and Foxconn and then spread to a number of other foreign-owned factories. A series of suicides occurred at Foxconn, protesting the low pay. The suicides at Foxconn received negative media coverage. Then, once again, in January 2012, about 150 Foxconn employees threatened to commit mass suicide to protest their work conditions.

It is tempting to think of these labor protests as isolated incidents that can be quelled by simply making the work conditions better and by increasing wages. And a number of employers did indeed take such steps to appease workers. However, the protests are a sign of a deeper, more fundamental, issue. Such worker discontent is only to be expected once readily available cheap labor becomes hard to find. An interesting statistic is the number of "mass incidents" of unrest, which grew from 8,700 in 1993 to 90,000 in 2010, according to the Chinese government. The government itself is well aware of the possibility of a systematic increase in such incidents. It has increased spending on domestic security quite consistently year-on-year.[22]

Lewisian turning point

The abundant availability of cheap labor in emerging countries such as China made it very attractive for many Western companies to move their manufacturing operations to such places. China became the factory floor of the world. This massive move of manufacturing to China elevated the living standards of millions of people living below the poverty line. With an increase in their living standards, laborers demanded higher wages. If their desire for higher wages and better work conditions went unheeded, workers resorted to strikes and protests. The labor unrest we just mentioned were a result of such discontent. As the wages increase, and the supply of cheap labor decreases, China will at some point hit the Lewis turning point.

The Lewis turning point, based on work by the economist Sir Arthur Lewis, is the simple idea that as labor becomes scarce, wage expectations pick up, and the profitability of business ventures declines.[23] In China's case, in the past the excess labor in the agricultural sector provided a steady supply of cheap labor for the industrial sector. There was no significant upward pressure on industrial wages because the agricultural-sector wages were low and laborers were content with the available wages in the industrial sector. Readily available cheap labor also kept inflation low. However, as agriculture surplus labor is exhausted, and living standards improve, there is a natural desire in the labor force for increased wages. As industrial wages rise quickly and profits are squeezed, investments fall. The labor unrest in China is related to the Lewis turning point and is a natural consequence of the shift of a developing economy from a low-productivity to a high-productivity growth model.

It is a fact that "made in China" is not so cheap anymore as labor costs have risen rapidly in the country. Average hourly wages hit $3.60 in 2016, spiking 64 percent from 2011, according to the market research firm Euromonitor. That's more than five times the hourly manufacturing wages in India, and is more on a par with countries such as Portugal and South Africa. The wage increase has translated to higher costs for companies with assembly lines in China. Some firms are now taking their business elsewhere, which also means China could start losing jobs to other developing countries where the costs are lower.[24]

Debate continues as to whether China has already cross the Lewis threshold, or whether that milestone is still in the future.[25] It has become increasingly clear, however, to both the government and the business leaders of China, that a new model is now warranted. Foxconn is embracing large-scale automation as a solution. The company plans to automate its Chinese factories using software and in-house robotics units known as Foxbots. While the name Foxbots sounds sci-fi-like, Foxconn can already produce over 10,000 Foxbots a year.

The company plans to carry out automation in three phases. The first phase involves automating jobs that are too dangerous or too repetitive for humans to perform. The second phase will fine-tune automation by improving efficiency and streamlining production lines. The final phase plans to automate entire factories and factory cities, employing only a minimum number of humans that are essential for the logistics and the inspection process. This is an ambitious plan. And its impact on jobs for humans will be significant. It remains to be seen how the government will react once automation affects a sizable number of jobs.

It is not just Foxconn that plans to automate jobs in order to solve the labor-related problems. According to the China Employer-Employee Survey, conducted by the Wuhan Institute, Chinese Academy of Social Sciences, Stanford University, and the HKUST Institute for Emerging Market Studies, which followed 1,200 Chinese employers and their 11,300 employees over a period of two years, labor shortages were among the employers' biggest worries.[26] Over 40 percent of these employers introduced automation technology. While manufacturing wages are lower in many other Asian countries, Chinese manufacturers are opting not to outsource, but to automate.

Will your money be managed by machines?

Job losses in Wall Street firms don't make the news because they happen regularly due to cost-cutting reasons. Job cuts on Wall Street also happen when business operations shut down in a unit or certain geography. However, it is certainly newsworthy when the job cut is happening for an altogether different reason.

In March 2017, it was widely covered by the business press and even by mainstream media that Blackrock, the world's biggest money manager, announced it would cut more than 40 jobs, replacing a few of its human portfolio managers with AI-based algorithms.[27] The CEO of Blackrock later clarified that the computerized trading is not linked with job cuts and that AI-based trading machines are not exactly replacing human portfolio managers. On the robot revolution, Larry Fink said that "We'll have the same amount of employees in our equity division a year from now than we do today."[28]

It should also be viewed in the context of passive vs. active trading strategies and trading experts highlighting that emotional balance and not intellect is the key to success in markets. Large banks and Wall Street firms are very quick in following the innovations of their competitors and hence, Blackrock may not be alone. Larry Fink at Blackrock is only following legendary fund managers such as Steve Cohen of Point 72, Ray Dalio at Bridgewater and Paul Tudor Jones who also believed that there are many tasks in fund management where machines have an advantage over humans. Interestingly, none of them are pioneers. Quant firms such as Two Sigma and Renaissance Technologies have been using AI for years to forecast the markets better and handle operations.

It is a warning sign and confirms the fears that AI is a genuine threat to even high-end jobs, but this is hardly surprising. Research has confirmed that active stock picking mostly underperforms the indices and broader markets and this makes customers uncomfortable with the high rates of management fees. The asset management firms are under pressure to bring down the costs and pass on those benefits to customers if they want to retain them.

For example, though Assets Under Management (AUMs) have increased for institutional fund managers,[29] there is increased scrutiny on performance. Clients are making it abundantly clear that they won't pay exorbitant fees if they don't see value. Increasingly, it is not just the fund management jobs that have started to look vulnerable, but it has also become much more difficult to forecast which jobs can be called safe.

Millions of workers will face challenges

Measuring the impact of AI on jobs is a complicated exercise and answering such a complex question is fraught with difficulties. However, there are some estimates available. Oxford University's Carl Frey and Michael Osbourne analyzed the potential near-future impact of the latest developments in AI and machine learning on the US employment market.[30] Using the up-to-date version of the *Dictionary of Occupational Titles* (DoT), the so-called 0*NET database, they categorized jobs as per their susceptibility to automation. A recent White House study carried out by the National Science and Technology Council also looked at this problem.[31]

Both the reports point in the same direction: the accelerated pace with which the fascinating breakthroughs in AI are making non-routine jobs susceptible to automation, doing away with the need for humans. According to Frey and Osbourne, about 47 percent of US employment is at risk of getting automated at some point in the future.

As we discussed, the jobs most likely to be impacted in the near term are low-skilled jobs held by people with a low level of education. The Organisation for Economic Cooperation and Development (OECD) has estimated that 44 percent of American workers holding less than a high school degree work in jobs made up of "highly-automatable" tasks, while just 1 percent of Americans holding a bachelor's degree or higher hold such jobs.

The White House Council of Economic Advisors (CEA) ranked occupations by wages and found that 83 percent of jobs making less than $20 per hour would come under pressure from automation, as compared to 31 percent of jobs making between $20 and $40 per hour and 4 percent of jobs making above $40 per hour. Autonomous vehicles—like the self-driving cars that are already being road tested—could dramatically alter between 2.2 and 3.1 million full-time and part-time US jobs, the CEA has estimated. But not all driving jobs are equally susceptible. For instance, school bus drivers—responsible for both driving and minding children—are unlikely to see their jobs disappear.

In short, while a quantitative measure of the impact of AI on jobs is approximate and uncertain, it is easy to see a significant impact. The impact will be different for developed and for developing countries. Developing countries, where a larger share of the population is employed in low-wage roles, will probably feel a bigger brunt. Governments in both developed and developing nations must carefully think of the consequences and find solutions to handle the aftermath of automation.

There is hardly any doubt that how AI will impact us could turn out to be significantly different to the impact that other technologies had in the past. As we discussed in Chapter 2, AI is different than other technologies (which were largely enablers and supports) because AI includes intelligence and computers take decisions and solve problems. AI is also not a people-intensive industry. This has serious implications for the impact of AI on jobs.

The new economic paradigm will also mean that the returns on education will diminish. Even if you increase the number of places in universities, opportunities are like a pyramid with limited room at the top. So, college education is what you will need more and more because almost everyone else will have such an education. But, whether it helps the way it has to date is not certain.

Notes

1 Stephen Hawking, "This is the most dangerous time in our planet", *The Guardian*, Dec. 1, 2016. www.theguardian.com/commentisfree/2016/dec/01/stephen-hawking-dangerous-time-planet-inequality (Accessed on November 22, 2017)
2 Elon Musk, World Government Summit, Dubai, 2017.
3 Kevin J. Delaney, "The robot that takes your job should pay taxes, says Bill Gates", *Quartz*, Feb. 17, 2017.
4 "Are robots hurting job growth?", *CBS News*, Sept. 8, 2013.
5 Claire Cain Miller, "Long-term jobs killer is not China. It's automation", *New York Times*, Dec. 21, 2016.
6 Bernard Marr, "The rise of thinking machines: How IBM's Watson takes on the world", *Forbes*, Jan. 16, 2016. www.forbes.com/sites/bernardmarr/2016/01/06/the-rise-of-thinking-machines-how-ibms-watson-takes-on-the-world/#18f6fc921e43 (Accessed on November 22, 2017)
7 www.livemint.com/Technology/6FNNbCiHKUfbDh7Lrs8XcM/How-firms-are-embracing-artificial-intelligencepowered-chat.html (Accessed on November 22, 2017)

8 http://media-publications.bcg.com/BCG-CII-India-Future-of-Jobs-Mar-2017.pdf (Accessed on 17 April 2018), www.business-standard.com/article/news-ians/india-needs-to-create-12-15-mn-non-agricultural-jobs-per-year-117031400902_1.html (Accessed on April 17, 2018)
9 www.censusindia.gov.in/vital_statistics/srs_report/9chap%202%20-%202011.pdf (Accessed on 17 April 2018)
10 http://hdr.undp.org/sites/default/files/rhdr2016-full-report-final-version1.pdf (Accessed on April 17, 2018)
11 www.businesstoday.in/current/economy-politics/indian-railways-jobs-applications-for-90000-positions/story/273688.html (Accessed on April 18, 2018)
12 www.livemint.com/Industry/0lESzufshJGU8Xy8cUanEK/Indian-IT-jobs-decline-in-first-half-of-FY18.html (Accessed on April 18, 2018)
13 www.business-standard.com/article/companies/automation-to-impact-indian-jobs-the-most-infosys-ceo-vishal-sikka-117021300203_1.html (Accessed on April 18, 2018)
14 https://yourstory.com/2018/04/indian-bears-brunt-trumps-visa-reforms-h1b-filings-time-low/ (Accessed on April 18, 2018)
15 www.facilities.aicte-india.org/dashboard/pages/angulardashboard.php#!/graphs (Accessed on April 18, 2018)
16 www.firstpost.com/business/indian-it-companies-cartelising-to-keep-freshers-salary-low-resulting-in-high-attrition-mohandas-pai-4277295.html (Accessed on April 18, 2018)
17 www.livemint.com/Politics/BphkOxYuir6OaYcTrBtldJ/AICTE-to-cut-number-of-engineering-college-seats-by-600000.html (Accessed on April 18, 2018)
18 www.dnaindia.com/academy/report-9-in-10-mbas-engineers-unemployable-2273475 (Accessed on April 18, 2018)
19 www.firstpost.com/business/indian-it-companies-cartelising-to-keep-freshers-salary-low-resulting-in-high-attrition-mohandas-pai-4277295.html (Accessed on 18 April 2018)
20 https://qz.com/445500/the-cost-of-getting-a-decent-education-in-india-is-now-staggering/ (Accessed on April 18, 2018)
21 David Barboza, "How China built iPhone City with billions in perks for Apple's partner", *New York Times*, Dec. 29, 2016. www.nytimes.com/2016/12/29/technology/apple-iphone-china-foxconn.html (Accessed on November 22, 2017)
22 Ben Blanchard and John Ruwitch, "China hikes defense budget, to spend more on internal security", *Reuters*, March 4, 2013. www.reuters.com/article/us-china-parliament-defence/china-hikes-defense-budget-to-spend-more-on-internal-security-idUSBRE92403620130305 (Accessed on November 22, 2017)
23 www.economist.com/blogs/freeexchange/2013/01/growth-and-china (Accessed on April 11, 2018)
24 www.cnbc.com/2017/02/27/chinese-wages-rise-made-in-china-isnt-so-cheap-anymore.html (Accessed on April 18, 2018)
25 Xiaobo Zhang, Jin Yang, Shenglin Wang, 2010. "China has reached the Lewis turning point". IFPRI Discussion Paper 000977. Development Strategy and Governance Division at International Food Policy Research Institute. www.ifpri.org/publication/china-has-reached-lewis-turning-point, http://ebrary.ifpri.org/utils/getfile/collection/p15738coll2/id/1737/filename/1738.pdf (Accessed on April 18, 2018); Mitali Das and Papa N'Diaye, 2013. "Chronicle of a decline foretold: Has China reached the Lewis turning point?". International Monetary Fund Working Paper. www.imf.org/en/Publications/WP/Issues/2016/12/31/Chronicle-of-a-Decline-Foretold-Has-China-Reached-the-Lewis-Turning-Point-40281 (Accessed on April 18, 2018)
26 "How are Chinese manufacturing firms coping with rising labor costs", Hong Kong University of Science and Technology, Institute for Emerging Market Studies, June 20, 2017.

27 Sarah Krouse, "BlackRock bets on robots to improve its stock picking", *Wall Street Journal*, March 28, 2017. www.wsj.com/articles/blackrock-bets-on-robots-to-improve-its-stock-picking-1490736002 (Accessed on November 22, 2017)
28 Matthew J. Belvedere, "We are not substituting stock-picking machines for humans, says BlackRock's Larry Fink", *CNBC*, April 6, 2017. www.cnbc.com/2017/04/06/we-are-not-substituting-stock-picking-machines-for-humans-says-blackrocks-larry-fink.html (Accessed on November 22, 2017)
29 www.willistowerswatson.com/en-IN/press/2017/10/Assets-of-worlds-largest-fund-managers-passes-US-80-trillion-for-the-first-time (Accessed on April 18, 2018)
30 Carl B. Frey and Michael A. Osborne, 2017. "The future of employment: How susceptible are jobs to computerization"? *Technological Forecasting and Social Change*, 114: 254–80.
31 White House Publication: *Preparing for the Future of AI*. https://obamawhitehouse.archives.gov/sites/default/files/whitehouse_files/microsites/ostp/NSTC/preparing_for_the_future_of_ai.pdf (Accessed on November 22, 2017)

4

THE ROLE OF CHEAP CAPITAL

> Human individuals and human organizations typically have preferences over resources that are not well represented by an "unbounded aggregative utility function". A human will typically not wager all her capital for a fifty-fifty chance of doubling it. A state will typically not risk losing all its territory for a ten percent chance of a tenfold expansion. The same need not hold for AIs. An AI might therefore be more likely to pursue a risky course of action that has some chance of giving it control of the world.
>
> —*Nick Bostrom*[1]

Masters of the Universe: Always searching for the right balance

If there was to be a title "Master of the Universe", many would agree that Albert Einstein would be one of the stronger contenders for it. Einstein's equations, originally published in 1915, truly expressed the nature of large-scale gravity for the first time. His equations could mathematically model the entire evolution of the universe. One crucial ingredient of his equations was a number called the *Cosmological Constant.*[2]

The cosmological constant controlled the rate of expansion of the universe. Add too little of it, and the universe would be doomed to collapse under the gravitational pull of its own weight and end in a singularity, the point where all laws of physics break down. Add too much, and the universe would be condemned to expand forever and inflate out towards a cold and empty fate. Einstein's theory pre-dated actual cosmological data by decades. In the absence of any data guiding his intuition, Einstein found a static universe, forever at equilibrium, more intellectually appealing than a universe that is inflating or deflating. With just the right amount of the cosmological constant, not too much or too little, the universe could stay at equilibrium, and avoid the two extreme fates.[3]

These notions of inflation, deflation, and ever so fragile equilibrium, reappear in financial markets and economics. Perhaps this is not all that surprising. Large complex systems, be it the entire universe, or the world economies, face similar issues when it comes to their evolution and stability. Instability, in the guise of inflation or deflation, always seems to lurk around the corner. Stability seemingly depends on a few adjustable numbers, sometimes of natural origin, and at other times of human creation. The cosmological constant postulated by Albert Einstein could balance the inflating and deflating tendencies of the universe. And, in the case of economics, it is the interest rate that can decide whether an economy attains the right amount of growth and inflation, or falters and stagnates into deflation.[4]

Up until the great financial crisis of 2008–09,[5] central bankers were widely considered as being privy to the secret recipe to always keep the markets in a state of well-being. Former Chairman of the US Federal Reserve, Alan Greenspan, was christened the "Master of the Universe"[6] or more precisely the "Master of the Financial Universe"; he was universally seen as the man responsible for having guided the US economy through the various and numerous vagaries of market forces. The economy, left to its own means, could veer off towards high inflation or depressing deflation, which could lead to recession, or a number of related sicknesses such as disinflation, stagnation, stagflation, and other such variations with negative impact on economy and people.

High inflation could drive up prices of all sorts of goods and services and lead to general societal unrest. Low inflation, or deflation, could be a sign of lack of growth and can create a depressing environment in society.[7] Central bankers, with their fingers on the dial of the interest rate, could tame inflation and usher in growth. Alan Greenspan exemplified this persona of the ever-powerful banker who kept the interest rates at just that magic level required to cure all malaises of the economy. Greenspan favored deregulation, where government regulators had little role to play, and low interest rates. During his tenure, the interest rate fell all the way from around 10% to 1%. The markets loved it. To be fair, who doesn't like cheap capital, where borrowing is easy, interest paid on borrowed money is low and markets are in a perpetual bull run.

Inflation targeting: What is too high and what is too low?

The interest rate is, of course, something that affects all of us. Banks lend money at a rate benchmarked to the interest rate set by the central bank. The mortgage rate depends on it, interest paid on student loans depends on it, and the return on certificate of deposit is derived from it. The interest rate determines the level of inflation in the economy, which in turn is related to price levels and growth.[8] If the interest rate is too low, it can lead to overheating in an economy. What this means is that a low interest rate makes it easy for people to borrow money. The cost of capital, which is the interest paid on the borrowed money, is then low. As it becomes easy to borrow money, investors freely make use of it to invest in all sorts of investments. This, more often than not, leads to too much investment in speculative undertakings.

Think of people borrowing beyond their means to finance multiple mortgages. Think of subprime lending. To put it rather simply, money then keeps pouring into all kinds of productive and unproductive usages as the hurdle rate for returns required on investment becomes lower and lower, and the economy overheats. Capital is misallocated. Speculation rises. Careful market participants are always wary of such situations. Prior to the financial crisis of 2009, central bankers were happy to leave interest rates at a low level. As interest rates stayed at lower levels for longer, speculative projects, best exemplified by subprime lending, gained steam. The economy overheated, and just like what happens in all other similar situations, the economy then crashed.

The other end of the spectrum is also problematic. If the interest rate is too high, it will be costly to pay for and service the borrowed money. Investments will fall. The economy will enter recession. This happened in the United States under the watch of the then Chairman of the Federal Reserve, Paul Volcker, in the early 1980s. The US was facing an especially severe bout of inflation when Volcker became the Fed Chair in 1979.[9] The cure was obvious: hike interest rates. Raising rates, however, is never popular politically. No one enjoys paying more interest on borrowed money, or paying more as adjustable mortgage rates move higher.

While Paul Volcker preceded Alan Greenspan as the Chairman of the Federal Reserve, their personalities could not have been more different.[10] Greenspan was outgoing and popular, while Volcker was withdrawn and almost stoic in his approach to the markets. Volcker took the unpopular step of raising the interest rate to an unprecedented 20% to fight inflation. And surely enough, the rate of inflation fell from a high of 14.8% in 1980 to below 3% in 1983. While the increase in the interest rate succeeded in killing the problem of inflation, the US economy promptly entered a phase of recession. As it became expensive to borrow money, investment plans were cancelled. And as the monthly interest on already borrowed money sky-rocketed, many businesses faltered and stuttered.

The right equilibrium is, therefore, hard to achieve. Central bankers try to keep inflation at a reasonable level. Not too high, and at the same time, not too low. This is called Inflation Targeting.[11] However, this is a hard game to play. When the dust settled and the world was counting the damages, both Volcker and Greenspan were found guilty of creating new problems, even as they solved old ones, as they pushed the economy to one side or the other of that happy medium in between. To be fair to both Volcker and Greenspan, no one knows where the balance lies and what is the right point of equilibrium.

The parallel we drew at the beginning of this chapter between the universe and the economy, between the cosmological constant and interest rates, between Einstein and central bankers, is now ripe for its natural conclusion. As observations by astronomers, led by Edwin Hubble, conclusively established an expanding universe,[12] presumably initiated by an event such as the Big Bang, Einstein came to grips with the absence of any equilibrium in the universe. The actual physical universe turned out to be not static, but expanding at a fixed rate. This prompted Einstein to call the cosmological constant the biggest blunder of his life.[13]

Alan Greenspan, the modern-day "Master of the Universe", the star central banker, similarly called his low interest rate policy and disposition towards deregulation a mistake. After the subprime crisis, at a Congressional hearing lawmakers asked Greenspan if he had been wrong. "Do you feel that your ideology pushed you to make decisions that you wish you had not made?" a lawmaker asked. "Yes, I have found a flaw. I don't know how significant or permanent it is. But I've been very distressed by that fact.".[14] Balance, it turns out, is a difficult thing to achieve. Both Einstein and Greenspan had the humility to eventually see the flaw in their thinking and where they had faltered. But, the fact remains that there was a flaw in their thinking.

Are there limits to the laws of money?

When it comes to large systems, physical or financial, striking that right balance between expansion and collapse, between inflation and deflation, can be difficult. Parameters such as the cosmological constant, or the interest rate, have a strong bearing on the stability of the system. The right choice of the cosmological constant can lead to the right conditions in the model of the universe that are conducive eventually for the evolution of life. And the right interest rate choice can lead to markets and the economy being conducive to economic growth and development.

Admittedly, this is merely an analogy between two very different systems. An analogy that is bound to break down once put under microscopic scrutiny, as most analogies do. However, let us take the analogy even further and ask what happens when the determinants of stability are driven to unusual values. Imagine a universe where the cosmological constant can change between different values. Start off with a small positive constant, and the universe is pleasantly expanding resembling ours. Now if the cosmological constant were to suddenly flip to a negative number, the entire universe would grow unstable, and collapse to a singularity, a point in space and time where all laws of physics break down.[15]

In an analogous way, the usual "laws of money" turn upside down when the interest rate goes negative in an economy. This might have seemed like a strange and esoteric possibility until just a few years ago, but not anymore. Over the last few years a number of countries, Denmark, Sweden, Switzerland, have pushed their interest rates to the negative territory.[16] They are hoping that as borrowers get rewarded for borrowing, they will spend more. And more spending will help the economy to grow out of the deflation that came about as a result of the subprime crisis. A number of other economies, from Japan to the US, have espoused a near-zero interest rate policy with the same hope of stimulating growth.

There is a lot that has been said about negative interest rates and how they were unthinkable even a few years ago. How negative interest rate policies have struggled wherever they have been implemented, and that they are unable to create the desired impact to reinvigorate economies. There are question marks as to how monetary policy has ceased to be a force that can take us to a nice equilibrium between growth and the cost of money.[17] In this context, recent developments also mean that some of the basic laws of money are breaking down. The concept of money in the conventional sense is inherently linked to some of the basic premises that underlie our financial intuition. These might no longer be valid today.

Let us talk about the "law of compounding" and the generation of new money with reinvestment. The more you start with and the sooner you begin, the better it is (of course, rate also plays a role but time and initial corpus are considered more important). The "law of compounding" has also been one of the most important driving forces behind the evolution and growth of modern finance. And in a scenario like today's when interest rates become negative, the "law of compounding" works in reverse. Or, a) the more money you have to start with, the more you lose in absolute terms, b) the more time you have with you, again the more you lose.

The possibility of negative interest rates was discussed in the early 20th century as an interesting theoretical possibility. Famously, the British economist John Maynard Keynes (or J M Keynes or simply, Keynes) supported the idea of a tax on money in one of his works in 1936 and believed it may help solve the problems of sluggish growth and unemployment.[18] However, he was also of the view that it might be unrealistic or only theoretical. But, coming to the present, in less than a hundred years, negative or zero interest rate based monetary policies have been implemented across countries in Europe and in Japan.

Is it not amusing that ZIRP (zero interest rate policy) and NIRP (negative interest rate policy) do not shock anyone anymore! One can argue that it is too early to pass a judgment on such unconventional monetary policies, but even the strongest proponents of these policies would agree that the impact has not been as immediate or as positive as originally expected. As a result, except for some select pockets, consumption and growth remain slow-moving. The more important problem is the next time central bankers (and for sure, there will be a next time sooner or later) need the ammunition of very low interest rates, they will have all but squandered it away.[19]

There are two strong philosophical points (not arguments yet due to lack of sufficient data) against zero or negative interest rates: a) there is a serious risk of capital misallocation in general because human nature is such that if something is made freely available, and there are no resource constraints, user discretion suffers and the allocation decisions become the first casualty. Japan might be immune because it is culturally very different, but that would not apply to every country following ZIRP or NIRP; b) savings are an equally important pillar for the advancement of humanity, and consumption alone cannot remain in focus for very long. The situation is very similar to rehabilitation therapy for drug addicts. Notwithstanding the withdrawal symptoms, drug dosage needs to be reduced and not increased.

Can low interest rates stimulate jobs?

It may sound like convoluted logic, but in an interesting 2013 paper, titled "Long-term investment, the cost of capital and the dividend and buyback puzzle", published in the *OECD (Organization for Economic Cooperation and Development) Journal on Financial Market Trends*, Adrian Bundell-Wignall and Caroline Roulet argued that low interest rates are bad for capital expenditure.[20] They argued that interest rates are so low today that they are not even enough to support banks. Hence, we are seeing a liquidity-driven speculative bubble in almost all the possible asset classes. They also said that since there is a disconnect between bond yields, equity risk premium and nominal GDP trends, it would be extremely difficult and volatile to raise interest rates in this environment.

More importantly, the authors looked at 4,000 global companies and highlighted that: a) capital expenditure decisions are influenced by cost of equity; b) buybacks are driven by the gap between cost of equity and debt; c) low interest rates work against long-term investment because debt finance is so cheap and cost of equity for making long-term investment (risky investment) may not have come down as much; d) when sales growth is low and uncertainty is high, companies would not incur capex (capital expenditure) but would look at inorganic growth as a preferred option; and, e) until we return to normal interest rates and reduce the incentive for low capex, it won't really help economic recovery.

Coming back to the present-day scenario, interest rates have declined further almost everywhere across the globe. And capex is still disappointingly low. Commodity prices have also collapsed over the last few years. A barrel of oil cost over $100 just three years ago, and now it hovers around $40.[21] Though some analysts try to link the low global capex with the weakness in commodities, the issues are interlinked. It is difficult to establish which one is the cause and which one is the effect. So, though it may sound counter-intuitive, cheap capital may have actually increased the incentive to postpone capex for companies. When the capital availability is abundant and the money is cheap, there is much more incentive for companies to buy ready-made capacity by going for inorganic growth through acquisitions as compared to incurring capex to increase capacity. And, most importantly, this is a question that is relevant for the excess froth in increased investment in start-ups and new ventures.[22]

In many industries across the world, there certainly exist issues such as: a) excess capacity and low end-user demand, and, b) numerous problems with many private companies and their lack of appetite for credit. Also, as clearly demonstrated in the paper "Long-term investment, the cost of capital and the dividend and buyback puzzle" mentioned in the first two paragraphs of this section, with a sufficiently large sample set of companies, low interest rates may not always be very helpful in reviving end-user demand and recovery. On the other hand, it is possible that a low interest rate may harm the capex plans and, in turn, the recovery. In a weak demand and poor sales growth scenario, no matter how cheap credit is, it is of no help.

Why is the "don't worry about losses" strategy of start-ups and e-commerce companies sensible?

Most established companies spend considerable resources on research and development (R&D) and innovation. Hence, they continuously focus on the development of new products. Even many of the "old economy" companies, which seem to be in an unremarkable and steady state on the surface, have some products that are emerging, some that are in the growth phase, some that are in maturity, and some others getting older and dying—simultaneously and all in one place. How do the companies measure the success of their "innovation" efforts?

For some of them, including industrial behemoths General Electric (GE), Siemens and ABB, it is very important to look at the revenue contribution of products on the basis of when they were launched. For example, a company may have a target of at least 25 percent of its revenue coming from the products that were launched in the last five years. Now, these so-called "new products" don't come from thin air; they are a result of massive investment in development, production and marketing. And, many (or rather most) of these fail in the marketplace, and this is despite very good intentions.

How does this new product pipeline get funded? The already-established products of the company that are "doing well" in the marketplace fund them. In a way, "cash cows" fund the future "stars". And, most of today's "cash cows" were "stars" once upon a time. Is this sensible? Yes, very much so. Unless you invest in the development of tomorrow's "stars" today, you will not have "cash cows" the day after tomorrow. It should not really matter whether this funding that goes into the development of new products comes from "internal" sources or "external" sources. In the case of many of the e-commerce companies, the funding comes from external sources.

Similarly, people should stop worrying about crazy valuations of e-commerce companies or start-ups. It is as simple as this: the worth of a product that is still in infancy is judged by management internally or by the investors outside the company. The money required to nurture a product that is still not fully developed comes from outside in the form of venture capital or PE money for start-ups. What is a reasonable valuation? No one knows really.

When a pharmaceutical company is investing millions of dollars in new drugs, it just doesn't know whether this much investment in a particular product is justified. When an automobile company is working on a new design of a car, it is not really sure if the buyers will really buy it. When you are producing a movie, you can never be sure about whether it will be a blockbuster or a dud. So you can never forecast your "Return on Investment" with certainty. The development of new products and new businesses is very similar.

The skeptics also say that most ventures will fail. Yes, they will. But, it is the same thing with any such pipeline of new products. Most of them fail and only a select few do succeed. There is one difference though. The way companies interact with customers has changed fundamentally with the advent of web and smart phones, so the probability that outsiders will succeed is much higher now. Theoretically speaking, it doesn't really matter whether products are getting developed with "internal accruals" or money being pumped in by "providers of risk capital".

So, stop worrying and enjoy the ride. It would be amazing to see how the change unfolds in front of your eyes. However, it would not be very difficult to argue that the relative importance (or cost or value, whichever way to define it) of "growth" and the "cost of growth in the form of capital needed for growth" matters. When you look at how the role of easy and cheap availability of capital has been instrumental, there is a clear conclusion that when too much money is chasing the same available "growth" opportunities, the valuations go through the roof.

Another way to look at the craze for start-ups is through equity markets' preference for growth. There are things that are important for investors when they analyze stocks such as including the earnings profile, pedigree of the company, cash flow generation and valuations. The investors look at many of them and other factors such as sector prospects, management quality & track record, auditors and corporate governance standard of the company before taking a investment decision.

However, everything else fades in front of "growth". Investors would not mind paying expensive valuations for companies that they hope will deliver superior growth. Profitability matters but growth matters more. Under usual circumstances and when everything else is the same, investors will prefer the company with "less profitability but higher growth" to a company with "higher profitability but lesser growth". This very well applies to VC investments too.

Cheap capital, factors of production, and is there a linkage with AI?

All over the world, labor-saving technologies and cheaper capital are putting economies at the risk of jobless growth. This is a situation where GDP growth rates and many other macroeconomic indicators may all look good but there would not be too many new jobs getting created in the economy. There is growth in the economy but there are not too many people actually benefiting from it.

Over the past 30 years, Asia, especially China, became the manufacturing hub and jobs started to disappear in the United States and Western Europe. However, now there is a real risk that countries will not be able to follow the path of industrialization to prosperity. The primary reason is cheaper capital that has made the labor cost advantage less relevant.

Manufacturing processes are more automated today than ever and this trend is unlikely to reverse. When a corporation is evaluating investment options in automated technologies, the cost of capital makes a huge difference. This does not just happen in more expensive manufacturing locations but equally impacts the cheaper-to-manufacture-at countries like China. For example, China is the fastest-growing market for industrial robotics and it has replaced the US as the biggest market for automation.[23]

And this is neither unexpected nor surprising. As the wage costs increased in China and other lower-cost locations such as Vietnam, Indonesia, Bangladesh, India and the Philippines became viable options, workforce automation is the only option left for the Chinese political leadership to maintain the country's competitive advantage. The declining costs of automation are driven by cheaper capital and better technology. And, job creation suffers with this trend.

The technologies of the 20th century helped cross-border production and in shifting manufacturing to cheaper locations. Developments in robotics, automation and AI can remove the need for outsourcing. With more automation, companies have all the reason and incentive to shift manufacturing to automated factories. It does not matter if we are talking about high-wage locations or cheaper alternatives in terms of manufacturing locations, workers will find it increasingly difficult to compete with cheaper technologies.

If corporations want to eliminate unskilled or semiskilled labor—like assembly-line workers, packagers, fast food workers—and the corporations are reasonably sure about the steady demand for the product, then automation becomes cheaper than unskilled or semiskilled labor. The investment decisions are based on Net Present Value (NPV). NPV is the sum total of the present value of cost savings minus the investment required today. When the cost of capital is cheap today, making investments becomes less prohibitive.

NPV is calculated by discounting future cash flows linked with savings by a certain rate that reflects the cost of borrowing money in the economy and then deducting the initial investment cost (which is the cost of buying machinery) from this.[24] This helps to arrive at the value of a project in today's economic terms. A particular project can also be compared with other projects to see which one is a better option.

When the capital is costlier, hiring more labor is usually cheaper since automation involves an up-front investment that is called the "fixed cost" or "capital expenditure". While the value of today's investment remains the same in calculation, a higher cost of capital reduces the value of future incoming cash flows. However when the cost of capital is lower, the NPV may become positive even for huge investments. If the cost of borrowing (or the cost of money) is very low in an economy, the present value of savings made possible by increased use of automation increases in quantum when we are doing the NPV calculation.

Another key factor is cost of labor involved. The cost of capital argument works for both low-skilled and high-skilled jobs. If a high-skill job needs to be replaced with AI, the investment required is more to develop a suitable algorithm and machine, but at the same time, the savings are also massive if the attempts are successful. Automating the high-skill jobs require more investment but they save more costs as well. In fact, a cheaper cost of capital may also increase the availability of capital if the corporations are willing to pay a little more premium over the prevalent cost of capital.

This is not the case with manufacturing alone. Earlier, the shift from manufacturing took people towards services. However, many jobs in the service sector are equally automatable today. The labor cost is increasing while the trend for automation cost is down. When the cost of borrowing money is at record lows, it's less inexpensive to replace a worker with AI.

When everything else remains the same, only the low cost of capital is a strong enough reason to transfer more and more work to machines. And this is not the first time either, as we have already seen this in many fields. Once, agriculture was extremely labor intensive. Today, agriculture production has increased manifold, but virtually everything agricultural is done by machines.

Faster growth of AI: Unintended consequence of cheap capital?

The moral of the story is, even with a reasonable interest rate, balancing can be a difficult act. Greenspan and Volcker both struggled. But we are in a whole new regime of ZIRPs and NIRPs now. AI proliferation could be an unintended consequence of the sub-zero cost of borrowing. At the very least, it is greatly influenced by the low cost of capital.

Most AI research work is linked to practical applications and industrial usage. In a way, it is a chicken and egg story. More willingness to adopt a newer technology by industry will attract more and better research in that field, and similarly, more research will initiate greater adoption by industry. Capital that was cheap and was getting increasingly cheaper was (and still is, and is likely to remain this way in the immediately foreseeable future) an important game changer in enhancing the attractiveness of automation and AI-based technologies for the industry. When you are supplying industry with capital at virtually no cost, any smart CEO and the Board of the corporation will be able to roll out more machines that can replace humans and automate the processes.

The choice of labor over capital or capital over labor has the cost of either as the most important factor. Cheap capital does two things simultaneously: a) it replaces labor, and, b) it also limits the increase in cost of labor on a proportionate basis. AI is replacing labor and cheap capital is making that process faster. We are not saying that automation and AI would not have grown in the last 20 years had capital been more expensive. But, we are reasonably certain that it would not have grown at the rate that it has.

It is interesting that two of the most widely cited books on the topic, *The Second Machine Age*[25] and *Rise of the Robots*,[26] were both written as a result of the Great Recession. As jobs refused to pick up, these authors took a deep, hard look around them, and developed the ideas. In this chapter, we have tried to show that QE, a consequence of the Great Recession, might itself be hastening the arrival of AI.

Notes

1 http://publicism.info/philosophy/superintelligence/6.html (Accessed on April 11, 2018)
2 "What is a cosmological Constant", Wilkinson Microwave Anisotropy Probe (WMAP), NASA. https://map.gsfc.nasa.gov/universe/uni_accel.html (Accessed on November 22, 2017)
3 "What is the ultimate fate of the universe", Wilkinson Microwave Anisotropy Probe (WMAP), NASA. https://map.gsfc.nasa.gov/universe/uni_fate.html (Accessed on November 22, 2017)
4 "How does monetary policy influence inflation and employment", FAQs, Board of Governors of the Federal Reserve System. www.federalreserve.gov/faqs/money_12856.htm (Accessed on November 22, 2017)
5 "The origins of the financial crisis: Crash course", *The Economist*, Sept. 7, 2013. www.economist.com/news/schoolsbrief/21584534-effects-financial-crisis-are-still-being-felt-five-years-article (Accessed on November 22, 2017)
6 Christopher Farrell, "The lasting legacy of Alan Greenspan", *Bloomberg*, Nov. 16, 2001. www.bloomberg.com/news/articles/2001-11-15/the-lasting-legacy-of-alan-greenspan, "The committee to save the world", *Time Magazine*, Feb. 15, 1999. (Accessed on November 22, 2017).
7 Timothy Cogley, "What is the optimal rate of inflation", *FRBSF Economic Letter*, Number 97–27, Sept. 19, 1997. Federal Reserve Bank of San Francisco. Brian Motley, "Inflation and growth", *FRBSF Economic Letter*, Number 93–44, Dec 31, 1993. Federal Reserve Bank of San Francisco. See also, www.frbsf.org/education/publications/doctor-econ/1998/june/inflation-economic-growth (Accessed on November 22, 2017)
8 *Instruments of the Money Market*, Ch 1: "The money market", Federal Reserve Bank of Richmond. www.richmondfed.org/publications/research/special_reports/instruments_of_the_money_market (Accessed on November 22, 2017)
9 "The Volcker Recession: Who beat inflation", *The Economist*, March 31, 2010. www.economist.com/blogs/freeexchange/2010/03/volcker_recession (Accessed on 22 November 2017); David Kestenbaum, "How former Fed Chairman Paul Volcker tamed inflation – Maybe for good", *NPR*, Dec. 15, 2015. www.npr.org/2015/12/15/459871005/how-former-fed-chairman-paul-volcker-tamed-inflation-maybe-for-good (Accessed on November 22, 2017); William Silber, "How Volcker launched his attack on inflation", *Bloomberg*, Aug. 20, 2012. www.bloomberg.com/view/articles/2012-08-20/how-volcker-launched-his-attack-on-inflation (Accessed on November22, 2017)
10 David Marsh, "Volcker and Greenspan", *MarketWatch*, Nov. 17, 2008. www.marketwatch.com/story/volcker-and-greenspan-a-study-of-contrasts (Accessed on November 22, 2017)
11 Sarwat Jahan, "Inflation targeting: holding the line", IMF Publications, Mar 28, 2012. www.imf.org/external/pubs/ft/fandd/basics/target.htm (Accessed on November 22, 2017)
12 "Edwin Powell Hubble – The man who discovered the cosmos", Hubble Space Telescope. www.esa.int/About_Us/Welcome_to_ESA/ESA_history/Edwin_Hubble_The_man_who_discovered_the_Cosmos (Accessed on June 24, 2018)

13 George Gamow, 1970. *My World Line*. New York, NY: Viking Press, p. 44.

14 Andrew Clark and Jill Treanor, "Greenspan – I was wrong about the economy. Sort of.", *The Guardian*, Oct. 23, 2008. www.theguardian.com/business/2008/oct/24/economics-creditcrunch-federal-reserve-greenspan (Accessed on November 22, 2017)

15 See, for example, Anthony Aguirre, "Eternal inflation, past and future", in *Beyond the Big Bang: Competing Scenarios for an Eternal Universe* (2008), Ed. Rudy Vaas. Berlin: Springer. Also in https://arxiv.org/pdf/0712.0571.pdf, p. 17. (Accessed on November 22, 2017)

16 "Moody's: Negative interest rates in Switzerland, Denmark, Sweden are having unintended consequences, with Sweden most at risk of asset bubble", Global Credit Research, Moody's Investors Service, Mar. 16, 2016.

17 Andreas Jobst and Huidan Lin, "Negative Interest Rate Policy (NIRP): Implications for monetary transmission and bank profitability in the euro area", IMF Working Paper, International Monetary Fund, Aug. 2016. www.imf.org/external/pubs/ft/wp/2016/wp16172.pdf (Accessed on November 22, 2017). Rima A. Turk, "Negative interest rates: How big a challenge for large Danish and Swedish banks", IMF Working Paper, International Monetary Fund, Oct 2016. www.imf.org/external/pubs/ft/wp/2016/wp16198.pdf (Accessed on November 22, 2017). Naoyuki Yoshino and Farhad Taghizadeh-Hesary, Hiroaki Miyamoto, "The effectiveness of Japan's negative interest rate policy", ADBI Working Paper, Asian Development Bank Institute, Jan 2017. www.adb.org/sites/default/files/publication/225371/adbi-wp652.pdf (Accessed on November 22, 2017). Scott A. Mather, "Negative interest rate policies may be part of the problem", PIMCO, Feb. 2016. www.pimco.com/en-us/insights/viewpoints/viewpoints/negative-interest-rate-policies-may-be-part-of-the-problem/ (Accessed on November 22, 2017).

18 John M. Keynes, 1936. *The General Theory of Employment, Interest and Money*. London: McMillan.

19 Claudio Borio and Boris Hofmann, "Is monetary policy less effective when interest rates are persistently low", BIS Working Papers No 628, Monetary and Economic Department, Bank for International Settlements, April 2017. www.bis.org/publ/work628.pdf (Accessed on November 22, 2017). "Monetary policy after the crash: Controlling interest", *The Economist*, Sept. 21, 2013. www.economist.com/news/schools-brief/21586527-third-our-series-articles-financial-crisis-looks-unconventional (Accessed on November 22, 2017)

20 Adrian Bundell-Wignall and Caroline Roulet, 2013. "Long-term investment, the cost of capital and the dividend and buyback puzzle". *Journal on Financial Market Trends*, 2013(1). www.oecd.org/finance/Long-term-investment_CapitalCost-dividend-buyback.pdf (Accessed on November 22, 2017)

21 Elena Holodny, "Timeline: The tumultuous 155-year history of oil prices", *Business Insider*, Dec. 20, 2016. www.businessinsider.com/timeline-155-year-history-of-oil-prices-2016-12 (Accessed on November 22, 2017)

22 Joanna Glasner, "2017 VC market: U.S. Start up investment picks up in Q1", *Crunchbase News*, April 11, 2017. http://about.crunchbase.com/news/2017-vc-market-u-s-startup-investment-picks-q1/ (Accessed on November 22, 2017)

23 "World robotics report 2016", International Federation of Robotics. IFR Press Releases, Sept. 29, 2016. https://ifr.org/ifr-press-releases/news/world-robotics-report-2016 (Accessed on November 22, 2017)

24 https://corporatefinanceinstitute.com/resources/knowledge/valuation/net-present-value-npv/ (Accessed on April 11, 2018)

25 http://secondmachineage.com/ (accessed on April 11, 2018)

26 www.basicbooks.com/titles/martin-ford/rise-of-the-robots/9780465097531/ (accessed on April 11, 2018)

5

ETHICS OF AI AND THE NEED FOR REGULATION

> I cannot conceive of a personal God who would directly influence the actions of individuals, or would directly sit in judgment on creatures of his own creation. I cannot do this in spite of the fact that mechanistic causality has, to a certain extent, been placed in doubt by modern science. My religiosity consists in a humble admiration of the infinitely superior spirit that reveals itself in the little that we, with our weak and transitory understanding, can comprehend of reality. Morality is of the highest importance—but for us, not for God.
>
> —*Albert Einstein*[1]

Regulation may not be the right answer, but "No Regulation" is no answer

Throughout history, regulation has remained a controversial subject in almost all spheres. There are endless debates about its effectiveness and the right and optimal quantum of it. As a result, it is very difficult to build consensus on issues such as:

- Do we need regulation and if yes, where do we need it?
- How much regulation is enough and how much is counterproductive?
- How to ensure that participants comply and what works better: carrot or stick?
- Do we need regulation to regulate the regulator?
- What should be the contours of a regulatory framework in terms of its objectives, scope, structure, oversight and enforcement?
- What will be the dispute resolution and grievance redress mechanism?

However, despite these questions, and knowing very well that no regulation is perfect, it is difficult to make an argument that the world would be better off without regulation. An analogy from the corporate world is apt here. One can criticize or shower accolades on a Chief Executive Officer or the board of a company on the basis of their performance and their decision-making, leadership in difficult situations, and the ability to inspire the team when the chips are down. However, to praise or criticize someone who has never been in that position, and never became a CEO for his or her performance, is meaningless. One can of course claim that this person would have made a good CEO or a bad CEO if one knew the individual well enough, but one cannot do a performance evaluation for "the best CEO XYZ Inc. never had". Just like the exercise "the best Presidents the United States never had" makes for an interesting read but is of little tangible value, many of these "what if" simulations are meaningful only in the controlled settings of a laboratory.

The situation is quite similar when talking about regulation. One can always comment on the efficacy of a regulation in hindsight, on the basis of how well it worked. But how "no regulation" would have performed is hard to visualize with any degree of certainty. And an important example of this comes from the fast-changing and rapidly evolving world of finance. The financial crisis of 2008 led to a big debate on the role of regulations and whether the crisis could have been averted.

Lessons in regulation from the financial crisis of 2008[2]

The years leading to the 2007/08 financial crisis were similar. The field of finance was getting more and more exotic, highly mathematical and complicated. The enhanced computing capabilities and cutting-edge algorithms were also helping in developing trading strategies only a few could execute and even fewer could understand. There was a continuous barrage of advanced quant models, risk models, complex derivatives and financial assets based on less and less real tangible value. Even some of the junior traders were running positions worth billions of dollars. It was as if finally people had discovered the magic formula to create wealth out of thin air.

Everyone was happy. People were becoming filthy rich, corporations had an abundance of cheap capital, Fed was basking in the glory of a long and uninterrupted uptrend in the economy, markets were in a perennial bull run, and the law makers really had no big reasons to feel concerned.

As long as the music is playing, you've got to get up and dance: Chuck Prince

Charles Owen Prince III,[3] commonly known as Chuck Prince, is Citigroup's former chief executive. In 2007, immediately before the financial crisis, Prince, who was leading Citigroup then, said that his bank had not pulled back from making loans to provide funds for private equity deals. He described his company's situation as a major provider of financing for leveraged buyouts, "As long as the music is playing, you've got to get up and dance. We're still dancing."[4]

It is a different story that the music stopped a few months later. In November 2007, Prince resigned from his post as CEO of Citigroup due to the failing mortgage industry. He was replaced by Vikram Pandit as the CEO of Citigroup and by Robert Rubin as its Chairman. Still, Prince left with an exit bonus valued at $12.5 million, in addition to the $68 million he received in stock and options he had accumulated during his career and a generous pension and benefits.[5]

In the lack of tightly framed and vigorously enforced regulations, this led to an unprecedented disaster. It was not possible, theoretically or practically, for the government and the regulators to find out the arcane details of all the financial instruments, and the resulting entanglement of worldwide financial institutions. Most market players and market commentators did not think it was important to have firm regulations in place to prevent such an entanglement of financial institutions, or to even prevent the fallout from an imploding financial system.

No one really thought it was important, or even necessary, to find out how regulators would provide direction to the financial world. Then hitherto unthinkable events happened, Bear Stearns and Lehman Brothers collapsed, AIG and Freddie-Fannie had to be rescued, people lost their livelihoods. To salvage the situation, an unprecedented worldwide program of increased and cheap monetary easing had to be initiated.[6]

To be fair to the critics of excess regulation, if we look back at the financial crisis of 2008, there are people on the other side who think we already had too many regulations of the wrong kind and it is over-regulation that caused the crisis. For example, the 2001 banking regulation rule is often blamed for cramming banks with extremely harmful mortgage-backed securities that proved to be an albatross around the neck. To put it simply, regulations can play havoc with the definitions, risk weight of different asset types, what is allowed and what is not, and with the possibility of creating an artificial skew that, if left unchecked, may lead to snowballing and disastrous consequences.[7]

Another example is the regulation determining how much capital banks should hold. If it is too much and banks hold more capital than what is optimal, it will stymie growth. And if banks hold too little capital, they will spread themselves too thin.[8] Arguments can be difficult to make one way or the other. Even if we concede that excess or bad regulation caused the crisis, there is no way to prove that no regulation would have been any better. Ultimately, "skin in the game" for individuals is perhaps a more important risk-mitigating factor than anything else.

In 2008, the markets waited until the subprime crisis before warming up to regulations. Before that, the influential opinion makers and doyens of the financial world, including the ex-Chairman of the Fed, Alan Greenspan, actively pushed back against regulations and favored free market forces. Have we learned anything from that experience? If we see the situation from that perspective, is it really worth waiting until an "AI crisis" before realizing the need for government regulation?

As we write this book, progress in AI is exploding. There are parallels that can be drawn with the pre-financial-crisis world. People are getting rich. Start-ups are springing up left and right. The more complex the nature of a start-up's promised product, the greater typically is the venture capitalist's interest in the start-up. New statistical models are emerging; Big Data is the buzz word, machine learning methods are evolving almost on a daily basis. To sum it up, AI is the rage.[9]

Is it wise to let this accelerate without any regulations? One might object and say there is no need for oversight by regulators who don't understand AI and who would only hinder its development, growth and progress. However, can anyone deny the need for a sane voice at times; the need for thinking of the wider implications, and the need for the development of a better environment to stop potential misuse?

There will always be people, with or without domain expertise in AI, who will hold the view of this being a false alarm. Their view may be that the Doomsday scenarios, and the need for regulation and its possible impact, are not going to play out the way a regulator may plan for or a central authority may envisage. This may very well turn out to be true, and that may not be a bad thing after all. However, the question here is not who is right and who is wrong. The more pertinent issue is the involved cost and potential benefits. And the following two scenarios need to be judged on that basis.

AI regulator is in place

Any AI regulation has a real potential to slow down things a bit.[10] But other than that it is hard to see any big negative outcome from well-thought-out and properly enacted regulations. AI regulations should focus more on practical and industrial applications and less on academic research labs. The objective of any regulation should be to assess the impact of the final product introduced into mass markets, and not to hinder pure research.

In this context, one very important issue is how AI will impact human rights. The problem emanates from the fact that AI is not free of bias, which has been proven in multiple experiments; and the worst part is that because there is no human intervention and machines are expected to be fair (or rather assumed to be free of bias), the bias may get institutionalized. The AI algorithms and decision-making methodologies work by using data (accumulated by humans) through codes (written by humans) and by machines (developed by humans) and there are inherent (even when not deliberate) biases in them. Another related problem, which is a big human rights issue, is lack of transparency—as algorithms get more and more advanced, complex and self-correcting, do we understand then enough? If there is bias or not complete transparency, we come to another possible consequence: how are we going to fix the accountability if and when things go wrong? All these are important dimensions from the perspective of how human rights may be impacted because of AI.

On the balance, the following questions (and our answers) summarize our thoughts regarding creating AI regulations:

- Can regulation create artificial barriers in some areas of AI research and applications? YES
- Can regulation change market dynamics and unfairly promote inferior technology? UNLIKELY
- Would regulators dictate the terms for researchers and scientists? ONLY PARTIALLY
- Would regulators play a role in limiting practical applications or placing a framework around them? YES

If we inspect most of the objections regarding regulations, they are linked to a possible negative impact on how quickly AI research can reach from:

Idea ——> Laboratories ——> Factories ——>Masses

Poorly crafted AI regulations do have the potential to become a distraction. There is also the possibility of such regulations stifling the pace of progress in AI development. However, it is highly improbable for well-executed regulations to fundamentally alter the basic principles and the market dynamics in AI. The regulator will only be able to alter the direction in some places because market forces are simply too strong and powerful. Regulations can certainly be a heavy burden on a new fledgling field.

However, AI as a field is no longer in its infancy.[11] As it now matures and showers applications impacting all fields of society, it becomes important to have debates about proper regulations, just as it is appropriate to have similar debates regarding regulations in the field of finance or nanotechnology or biotechnology.

AI regulator is not in place

The absence of a regulator can bring about certain benefits. Most of these benefits are linked to the speed with which AI can transform the world. AI has been responsible for altering many industrial and business processes. It has even fundamentally changed the way humans and technology interact. However, any scientific or technological pure research is blind to its applications. The impact of a technology cannot be assessed by that piece of technology itself. It can be difficult for researchers to foresee wider consequences with their focus set on immediate priorities. On balance, the following questions (and our answers) summarize our thoughts towards the lack of AI regulations:

- Can "no regulation" create a more equal and democratic environment for research? YES
- Can "no regulation" take care of what is an acceptable area of research? NO

- Would "no regulation" be able to mitigate or slow down the negative impact on the masses? NO
- Would "no regulation" be able to handle the rogue elements in the field? NO

The function of a regulator is similar to the steering wheel and brakes in an automobile. Research is like the engine of an automobile. The brakes and steering wheel do not replace the engine, nor are they inconsistent with the engine's existence. The steering wheel is needed for course correction and for ensuring that the direction remains consistent with the path towards the destination. And, as we have pointed out in the first set of questions discussed at the beginning of this chapter, the objective is to ensure that a remedy exists when harm is a result of the use of the technologies. Regulation is also needed to answer broader questions about liability, etc. in the context of AI. Finally, brakes and discretion in how to use them are needed, to avoid accidents that can prove to be fatal.

Accidents still happen. There is no guarantee that a regulator will always be successful and that it will always be possible for regulations to remain ahead of practitioners. Certainly, a regulator's job will be even more difficult in a field like AI and the outcome more uncertain. But, this should not dissuade people from giving the need for regulations in AI a serious thought. On the basis of a simple straightforward cost-benefit analysis, having a regulator in place for AI is a bet worth taking.

Today, AI lacks a balanced and well-researched critical opinion

There are big divides in today's society, divides that lead to strife and disharmony. There is a divide between the rich and the poor, a divide between conservatives and liberals, a divide between city dwellers and the rural population, and many more. Another similar divide is between the technological literati and those who struggle to keep up with the fast-developing technological landscape.

This lack of technological awareness in a whole swathe of the population renders them lacking in knowledge about the direction and recent advances in science, in general, and computing, robotics and AI, in particular. Even people with a science and engineering background may not have the time and the capacity to keep up with the latest developments in very specialized fields such as convolution neural networks or reinforcement learning.

This hurdle that many people face in keeping up with the latest in AI means that not many people are even aware of the new AI developments, let alone the ethics of these developments. This includes not just people who are not immediately feeling the impact of AI, but also, a) people who would not be able to figure out the real cause behind these changes in their environment because they lack the required expertise and understanding to follow these advancements and, b) people who may have the expertise and required qualification to follow these developments but lack the motivation because they don't see the practical utility of investing time and effort in it.

Because of many contributing factors, some of the advanced scientific areas are hugely lacking a balanced analysis and viewpoints that look at different aspects in a more holistic manner. AI literature is in a pristine stage today. While it is technically challenging to keep up with AI developments, there is no dissemination of misinformation on AI, no propaganda against AI. One shudders to think of what could happen if the field of AI gets politicized. If the same data is viewed and the conclusions reached thereafter are different on the basis of one's subscribed ideology and depending on what one wants to believe, one can simply pick and choose. We see this unfortunate occurrence in many areas.

Climate change: Is it really happening, who is responsible and who should pay?

Let us look at a contrasting example in the topic of climate change. The science of climate change itself addresses a unique possibility in the sense that there are no two ways about it. Climate change is either happening or not happening. Climate change is either man-made or not. The answer itself is unique.[12] The answer should not depend on whether the seeker is from the right or the left side of the political spectrum, or whether the question is posed by the owner of a factory or a worker in the assembly line. However, various interest groups and the politicization of climate-change-related questions have made the field very confusing to the lay person. All sorts of noise bounce around in the media regarding climate change. This makes it very difficult to find out even the basics underlying climate science facts.

Climate change story 1: Why is Elon Musk more important than the leaders of 200 countries combined?

The Paris climate deal is back in the news after the US President, Donald Trump, decided to withdraw from the agreement on emission cuts that was signed about 18 months ago. However, our story is from a back date. It happened around December 2015 when the ink had not even dried on the "Historic Climate Change Deal in Paris" and reports quickly emerged about different interpretations of the potential impact. It is interesting how the coverage in developed countries and developing countries was vastly different on the implications of this deal.

We think it is more of a proof of the deal's ambiguity that there are no clear-cut "winners and losers" among countries. This is surprising because in the last 20-plus years, climate change talks have always been more about bickering between countries as the issues are contentious. One thing that most people would perhaps agree with is that the Paris deal was much more loosely worded than what was envisaged previously. Perhaps when so many countries gather for something as big as this, no one really wants to take the blame for the collapse of talks. So, one dilutes the provisions and postpones the bickering to a later date.

Considering that previous attempts on climate change have at best shown mixed results, this agreement ideally required having more clarity on "how", but in many cases the agreement did not go beyond "what". A few messages that could be deciphered from this deal were: a) the distinction between developed and developing countries was much less clear, b) the funding sources and sharing were not clearly defined, c) the peaking of emissions date was not specified and only talked about "as soon as", and, d) all the parts of the agreement were not legally binding.

Another interesting issue was that there would have been a continuous process of negotiations every few years; there was no guarantee that the commitment will remain when the agreement becomes less "subjective" and more "objective". Please note that we would not like to make it a comparison between developed and developing countries, but it was alleged that developing countries had given more than what they got in return.

This agreement is interesting to look at as an application of Game Theory inspired thinking. And, from this angle, there can be serious hurdles to its effectiveness. The only way fossil-fuel-based technologies can become obsolete is when cleaner technologies become easier to use and cheaper. Hence, in our view, Elon Musk is more important for the global climate than the leaders of 200 countries combined.

Climate change story 2: Solar is the new coal in India

In one of the recent auctions at the time of the writing of this book, India's solar power tariffs fell to a new low of Rs 2.62 per unit (approximately 4 cents per unit). This was for the auction of a 250MW capacity plant, and South Africa's Phelan Energy Group made this bid. The solar tariffs in India are coming down with each new bid. Just a few weeks ago, Solairedirect quoted a record-low tariff of Rs 3.15 per unit (approximately 5 cents per unit). This is an extremely interesting situation because concerns regarding the sustainability of these tariffs have not abated.

When we speak to energy sector experts, we find that most of them are worried that these tariffs are not viable. This is despite a sharp decline in installation costs. Indian renewable energy companies have been attracting huge interest from investors. This is despite solar tariffs falling incessantly. Indian renewable space is attracting huge investor interest and Indian renewable energy companies attracted over $1.62 billion during the first quarter of 2017. Funding raised by transactions in Indian solar and renewable energy companies was half of the total global funding raised by solar companies in the first three months of 2017.[13]

Most conventional plants have faced challenges with fuel and execution. Fuel is not a factor and execution challenges are also relatively benign in renewables. In terms of the Indian government's preference and support, renewables are clearly ahead vs. conventional sources of power generation. Earlier, the challenge was high cost but now solar tariffs are lower than coal-based tariffs because solar tariffs are now lower than average tariffs for conventional fuel-based power plants.

For example, NTPC (India's largest power utility with a 50GW-plus operating generation capacity and which is primarily a coal and gas-based generator) has average tariffs higher than 5 cents per unit. This means that more than 60 percent of India's generation capacity that is coal or gas based becomes commercially unviable as renewable electricity is cheaper now. It is important to note that no regulator or government initiative can achieve these changes unless they make sense commercially.

What happened with climate change should not happen with AI: Biotech lessons

The point we are trying to make with the above two stories on climate change is that market forces are immensely powerful. Even with the best of intentions and with the use of coercion (in terms of incentives and penalties), it is difficult to stop market forces. On the other hand, as we have also learned from the recent financial crisis, giving market forces free rein is not appropriate for every situation. Allowing market forces to run amok in each and every case is a recipe for disaster.

However, the overarching lesson for AI from climate change is not quite the role of market forces. It is, instead, the level and the quality of public knowledge and policy debate. Climate change discussion has become mired in politics. One hopes that something similar does not happen to AI. At present, notwithstanding the technical jargon, clear answers exist about the mathematical and technological basis of AI.

Areas of AI where debates are occurring are also accessible to those who care to follow the debates. One can clearly follow both sides of the argument. Until now, there do not seem to be any AI lobbyists or interest groups that are actively trying to tilt the public information one way or the other. It is important that no misinformation and rumors spread about a field as important as AI. The field of climate science is a clear warning. Misinformation and rumors can set a field back by years.

One positive example can be drawn from the field of biotechnology. Thirty years ago, biotechnology and genetics were at a similar stage to AI research. These fields showed enormous promise, and an ability to significantly alter the future course of human development. Scientists realized the importance of public education in these fields. There was an active effort to educate the public through workshops and media programs.[14] As a result, the public received undistorted information about biotechnology and genetics. This helped policy makers make informed decisions regarding the ethics and regulations in biotechnology.

AI is similarly a technically complicated field with an enormous possible impact on humans. There are complex questions when it comes to the ethics of AI.

Experts in the field need to ensure that correct and undistorted information is available for the public. This will get people to talk constructively and productively about various aspects of AI. AI is a game changer, the technology has immense potential, and at this stage we are only scratching the surface. But to make it work better, and to ensure it helps people make their lives better and more productive, we need informed opinion and a well-thought-out plan.

Who is talking about the ethics of AI?

While the layperson might lack the technical know-how, one wonders why AI experts were, until recently, not discussing actively the right approach to creating an ethical framework for AI, robotics and automation. The situation has changed a bit thanks to a handful of well-known personalities, such as Elon Musk, Sam Altman, Stephen Hawking, Bill Gates, and others, who have voiced concerns regarding the lack of discussion on relevant issues. Slowly and increasingly now, experts have started taking note of the hairy problems than can arise in the ethical dimension of AI.

Like any scientific and technological field, the technical problems littering the path of development in AI are complex and challenging. The amount of effort needed to solve these problems can be an all-consuming task for the experts. Just like any other field, such as nuclear power or genetics, it is only when the technological problems have been solved that the experts turn their attention towards the ethical questions. In the initial stages, the expert innovators, researchers and scientists may not have the bandwidth to tackle the ethical questions.

There is always a tension between the amount of technological progress that has to happen before the ethical issues begin to become relevant, and the pressing need to address ethical issues before it is late. What is the right point of time when a scientist must take pause from the ever-consuming technical research, and direct attention to ethics? At times the answer is only clear in retrospection.

Technological changes can occur swiftly. Recent progress in AI has, at times, even exceeded expert expectations. Eric Brynjolfsson and Andrew McAfee, authors of the popular and impactful book about evolving technologies and AI, *The Second Machine Age*, have written in the second edition that technological development has been much faster than they had imagined.[15] Between the publication of the first and second editions of the book, technological advances appeared on the horizon faster than even the most optimistic researchers had anticipated.

As technological progress becomes faster, sometimes it pays to be proactive in confronting the ethical and moral ramifications of the material progress. Especially if there is a possibility of an unforeseen risk that is hard to assess in its nature and magnitude, waiting for too long holds the possibility of exacerbating the negative effects. It is instructive to discuss two previous scientific advances that had clear ethical consequences, and that dealt with the moral and ethical topics in different ways—nuclear energy and biotechnology.

Nuclear energy: Is it a harbinger of peace despite Hiroshima and Nagasaki?

Theoretically imagining a scenario and witnessing it first hand are two quite different things. An average adult has likely watched hundreds of murders on TV shows and movies, but witnessing a homicide can have an entirely different impact on the mind of the witness. As another example, the scientists and technicians working on the development of the nuclear bomb as a part of the Manhattan project were not unaware of what they were creating and how it would be used.

But the nuclear test, and then the bombings of Hiroshima and Nagasaki, were still intensely shocking events for most of them.[16] They were not prepared for the practical consequences of the nuclear detonations. Imagining something theoretically, or in a laboratory setting, and seeing something with all its practical impact, can have entirely different connotations. We cannot put this in better words than Yogi Berra who famously said **"In theory there is no difference between theory and practice. In practice there is."**

In the case of nuclear energy, it can be argued that from 1945 onwards, the world has been, relatively speaking, a much more peaceful place compared to any other 80 odd years of recent human history. Therefore, was nuclear capability not a worthwhile technology to develop? Therein lies the paradox: nuclear technology has been destructive but it has also worked as an effective deterrent. Not everyone would agree to it but we think the argument is fairly strong.

Starting almost immediately after World War II and as a response to widespread massive destruction and loss of human lives, policy makers started to discuss ways to control proliferation of nuclear technology and responsible robust mechanisms to control the usage. The evolution and development of policy response towards nuclear technology was late, but nevertheless very effective. And it might be too early to tell, but it seems like this has worked well enough so far.

Biotechnology: The conflict with religion

In genetics, because of its sensitive nature and the potential for conflict with philosophical and religious ideas, the questions on ethics started to arise much earlier. In 1975 Paul Berg, a Stanford University researcher and leading scientist on DNA technology, organized the *Asilomar Conference on Recombinant DNA*.[17] The objective was to prevent deliberate and innocent damages that may be caused by DNA research and to develop a response to ethical issues that were at the forefront. The conference was important and effective in developing the thinking needed to tackle the issues involved.

And this has continued over the last 40 years. As recently as December 2015, an International Summit on Human Gene Editing took place in Washington under the guidance of biologist David Baltimore. Members from the United States, Britain and China discussed the ethics of germline modification. They agreed to support research under appropriate legal and ethical guidelines. For example, a specific distinction was made between somatic cells and germline cells. The effects of changes in somatic cells are limited to a single individual while in the case of germline cells, the genome changes could be passed on to future generations.[18]

There have been local controls as well. For example, British scientists were given permission by regulators to genetically modify human embryos by using CRISPR-Cas9.[19] However, researchers were forbidden from implanting the embryos and the embryos were to be destroyed after seven days. The United States has a regulatory system to evaluate new genetically modified foods and crops. The Agriculture Risk Protection Act of 2000 was an important step in stopping the spread of plant pests or noxious weeds to avoid unpleasant outcomes.[20] The act regulates any genetically modified organism that utilizes the genome of a predefined "plant pest" or any plant not previously categorized. Such regulatory interventions have played an important role in controlling the undesirable effects that may result even inadvertently.

Coming back to automation and AI, fields that arguably have a wider potential reach than nuclear science or biotechnology, the need for clear guidelines for the future direction of research and development is palpable. Technology with such vast implications for employment and leisure, for experimentation and entertainment, for war and peace, and, perhaps, for the existence of humanity, clearly should lean on a clear set of ethical and moral guidelines.

What should AI be allowed to do, and what should be off limits? Who should monitor the behavior of autonomous systems in fields such as medicine or finance? What level of integration between AI and humans is OK? Should autonomous systems be permitted in financial markets and if/how would they be penalized if these systems lead to mishaps, for example, a flash crash in the stock markets? What about the free market? How to divide AI regulation responsibilities between corporate and government bodies?

Even if we succeed in clearly enunciating a set of guidelines for AI research and development, should they be strictly enforced? Who should enforce them? Should there be a worldwide regulatory body to address disagreements between nations? Think of the disagreements about the nuclear nonproliferation treaty or the Paris climate change accord. The first few significant steps in the right direction were taken by the Future of Life Institute's Beneficial AI Conference in 2017[21] discussed in the next chapter.

The ethics issues may not be entirely straightforward

Local vs. global values

To see the kind of conundrums that arise with AI ethics, let us go back to Deep Blue, the chess program that beat Gary Kasparov. The programmers who made Deep Blue did not tell it how to act under every single situation. It had an overall goal: to win the game. In order to win the game, Deep Blue was given smaller sub-goals. However, when it came to detailed chess moves, Deep Blue was left to learn and evolve by itself. Likewise, AI can be told how to act, for a specific task or set of tasks, and the overall objective. But when it comes to each single step involved, AI, depending on how it was designed, can be on its own.

Many of the learning algorithms, such as types of neural networks and genetic algorithms, can evolve in non-transparent ways in order to achieve the final goal. It is impractical to pre-program every AI in such a way that each of its individual actions conforms to a predetermined set of values and ethics. In other words, while it might be possible to ensure that the global behavior of AI conforms to a set of values, if we want AI to learn and evolve in a non-trivially intelligent fashion—in a human-like fashion—then it might be impossible to make the AI's local behavior adhere to those sets of values.

Compare this with how humans have evolved as individuals and as a society. Humans have some quite basic sets of needs and some evolved sets of needs. Humans need to eat, reproduce, etc. Many of the actions taken by humans, even if they achieve clear goals in the short term, turned out to be harmful for nature. We have left indelible imprints on nature. How do we ensure that AI, which can have a much faster and bigger impact, does not repeat the same mistakes?

Transhumanism

Another interesting idea discussed by Bostrom is that of transhumanism.[22] There are already start-up companies in Silicon Valley that plan to develop AI systems that can be inserted into human beings. A simple example would involve your computer. The keyboard is an interface between the computer's hard drive and you. Typing is a slow process. So, while the computer itself acts very fast after receiving your instruction, sending the instruction to the computer via the keyboard is a slow process.

What if there was a way to construct a keyboard that is a part of us? What if we could instruct a computer using our thought process? One can see where this path would lead us—hybrids between computers and humans. This is uncharted territory. Is it acceptable to enhance human capabilities in this way? We have resisted biotechnology research with similar long-term, possibly hereditary, consequences. The moral and ethical issues that arise when talking about transhumanism are especially sticky.

Pricing manipulation at online retailers: It is legal but is it ethical?

The American retail industry is undergoing a crisis. Retail stores are closing almost on a daily basis and in large numbers. It is not that people have stopped buying things, but the shift to online purchase has proved to be the "kiss of death" for brick and mortar stores. However, there are subtle details to this issue that were not understood and appreciated in the past.

1. Online retailers have data about customers that has been collected over the years. On the basis of data collected on our online purchase habits, they can manipulate many things but most important among them is pricing. It is as simple as this: the price we will be asked to pay is a complex equation with numerous variables and the price we are going to pay will depend on our cumulative purchase behavior online.
2. Let us accept it. Apart from convenience, or rather even after including convenience, the most important reason why online retail picked up so quickly was—competitive pricing. The buyers realized that they can get the best deals online and that the pricing online will be more competitive compared to brick and mortar stores. The assumption was that since online stores can't attract buyers on the basis of their location, décor and available choices, they will offer customers better pricing.

As discussed in the recent article "How online shopping makes suckers of us all" by Jerry Useem, published in *The Atlantic*,[23] online retailers have already started using highly sophisticated ways to adjust and personalize prices to maximize profits. The article talks in great detail about how the price that will extract the most profit from a consumer has become a hot topic for a large number of researchers. The so-called "price experiments" have become a routine exercise because the right price can change by the day or even by the hour. Officially, though, not many would admit this. For example, Amazon says its price changes are not attempts to gather data on customers' spending habits, but rather to give shoppers the lowest price out there.

The article raises a fundamental question: Could the internet, whose transparency was supposed to empower consumers, be doing just the opposite?

When online merchants have the tools to dynamically adjust prices based on factors like a) where the customer is from, and b) what the customers have been buying and at what price and how regularly, then online retailers have the power to create a demand-supply equation for an individual customer. Economics 101 has taught us the demand-supply equilibrium for the market as a whole, but online retailers are taking it to an altogether higher level when the demand-supply equilibrium is not for the market but for the individual!

The implications are very interesting or threatening or frightening, depending on how you look at them. More granular data and consequent pricing decisions are good for online merchants and they can maximize profits and attain higher profits. But, are they collectively also good and fair for customers? Look at it from another perspective: your neighborhood store always knew which brand of soft drink you like but it never charged you a higher price for that on the basis of the information it had. But, online retailers have the ability to do this and most people would not even find out.

Not every person expresses the same demand for a product; different people have different levels of need for the same product. The enormous amount of data online retailers have empowers them to dissect customer buying patterns and then effectively guide the individual's demand. There is no market equilibrium price in this new scenario; it is an equilibrium price for individuals.

Scenario 1: How much will you pay for an air ticket?

Imagine you have an emergency in the family and you have to travel from Chicago to Boston and need to start as early as you can. The airline companies have access to all your data (phone calls, your WhatsApp messages, your emails, your Facebook posts). Now, the sophisticated and complex algorithms at the servers of these airlines can quote a fare to you on the basis of your desperate need to travel. And this is not the same as expensive fares during holiday season because during holiday season, the market equilibrium itself shifts to a higher price because of higher demand. The result is: more revenues and higher profits for the airlines.

Scenario 2: What is the right price for a book you need?

Imagine you have just enrolled in college and you desperately need that Advanced Mathematics textbook because it is a part of the prescribed texts for your coursework. You didn't have this need six months ago and you won't have this need six months later. Hence, your willingness to pay a higher price for this book is at the maximum today compared to any other time in history or in the future. Online retailers can manipulate prices if they know, on the basis of your recent searches, that you are in a desperate need for this book today. They are likely to quote a higher price to you and you would still buy the book. The result is: more revenues and higher profits for online retailers.

Free market principles and ethics of business: The broader question

As of now, there are no laws that prohibit online merchants from making such pricing changes. But, do we need pricing controls to stop such dynamic pricing on the basis of an individual's need and purchase history or are the online retailers simply trying to eliminate the inefficiency in the conventional model that doesn't allow individual pricing? We believe that most people may want the law to interfere to stop such pricing discrimination or exploitation of customer data for the merchant's gains. But would that count as tampering with free market principles[24] and the basic tenet of demand-supply law.[25]

The broader question is not just about the existing law, but about morals and ethics. And based on these parameters, the behavior of some online retailers would certainly fall in an ethically ambiguous zone. Today we are only scratching the surface in terms of what is possible with better algorithms and advanced AI tools. Imagine if AI becomes so powerful that by using it, online merchants can quote a price that is different for everyone and also different at all points of time: what would our reaction be? The question we need to collectively ask is whether we can leave everything to free market principles? AI is just too significant to take that chance perhaps.

AI: The legal and ethical dilemmas

In this context, it would be useful to examine the different ethical and legal dilemmas posed by AI. The subject is vast and there are several things that could even be unimaginable at this stage, but we think it will help if we can look at critical questions, including the ones related to Control, Human Rights, Dignity of an Individual, Privacy, Security, Intellectual Property, Transparency and Liability. Honestly, we fully understand the limitations in terms of how much can be covered here but the idea is not to have a comprehensive discussion on these issues but to at least begin with it so that we can be more sensitive about the legal and ethical dilemmas AI could lead to.

1. **Control**: In AI, the control problem is the hypothetical puzzle of how to build a superintelligent machine that will aid its creators, but at the same time avoid inadvertently building a super intelligence that will harm its creators. Its study is motivated by the claim that the human race will have to get the control problem right "the first time", as an evil AI might decide to "take over the world".[26] This is one of the key questions when we talk about the existential threat from AI.
2. **Human Rights**: The AI discussions are mostly currently limited to immediate, tangible impact but there are also challenges with the potential impact on human rights for machine vis-à-vis humans. In October 2017, Saudi Arabia became the first country in the world to give a robot citizenship. The robot, Sophia, has declared that she wants to use her unique position to fight for women's rights in this Gulf nation where the laws allowing women to drive were only passed last year and where a multitude of oppressive rules are still actively enforced (such as women still requiring a male guardian to make financial and legal decisions). Sophia seems to have more rights than women living in Saudi Arabia.[27] So far, at least theoretically, we have been working with the premise that humans are superior to other living and non-living things. Will this premise change when AI becomes intellectually superior to the human race? There are a lot of questions posed by the entry of powerful and intelligent AI machines and the most important one among them is: does AI pose a threat to human rights?[28] In fact, some of the researchers are also claiming that AI machines deserve human rights.

3. **Bias**: There is also a risk that AI could lead to biased behavior, as bad as or even worse than humans. Machine learning based on human behavior has a risk that it may be transferring the historical biases in our society to machines. This could mean, for example, that AI used in predictive policing or loan-approval systems would entrench discrimination on the grounds of race or gender.[29] AI can be used for profiling of a population, and that could lead to undesirable results that will not have any basis other than simple bias of the machine.
4. **Privacy**: There is also a significant danger that the personal data of a large population retained by machines will be accessed for unethical and illegal purposes. These attacks risk the right to privacy of users.[30] When it is serious, it could lead to loss of identity, loss of property, or even loss of life. The risk could also apply to human dignity in cases where there is a sensitive information leak. AI could inadvertently lead to more of these issues.
5. **Intellectual Property (IP)**: AI development is expensive and there are several areas in which multiple companies and agencies are working and competing. For some, it is a mission and for many others, it is business. When there is a commercial angle to it, it is likely that there could also be instances of disputes and controversies linked to who should be given credit for a certain development and how to protect the intellectual property rights of developers and companies. As we have seen in the case of data, this is an important issue.
6. **Transparency**: When AI is getting better on its own and self-correcting behavior is an essential characteristic of these machines, will it always be possible to track all these changes and get a clearer picture of when and how a particular trait became a part of AI machines? We already know that machine learning is leading to automatic advancements in the algorithms and we might not know why something got changed?
7. **Liability**: The part that is linked to how to fix the responsibility if something goes wrong is no less tricky. If there is an issue with data security and privacy and also when there is less than optimal transparency on which process should be attributed to whom and which phase of development was responsible for which decision, accountability will suffer. And of course, the liability issues get more complicated when there is less clarity on accountability.

Use of AI for profiling: How fair is the discrimination?

The *Cambridge Dictionary* defines Stereotype as "a set idea that people have about what someone or something is like, especially an idea that is wrong" or "to have a set idea about what a particular type of person is like, especially an idea that is wrong". The most common stereotypes are sexual and racial".[31] In information science, profiling refers to the process of construction and application of user profiles generated by computerized data analysis. This involves the use of algorithms or other mathematical techniques that allow the discovery of patterns or correlations in large quantities of data, aggregated in databases. When these patterns or correlations are used to identify or represent people, they can be called profiles.[32]

Profiling is a type of pattern recognition and hence, it enables extensive social sorting. And that is where the problem is. Some of the areas where AI could lead to serious social tensions and cornering of resources (both economic and social) by an already well-off population could be as follows:

1. **Hiring**: It's not just humans, computers can be prejudiced too as software may accidentally sort job applications based on race, age or gender.[33] As AI algorithms get more sophisticated, the bias may become more difficult to catch but will there be attempts to eliminate the bias or will it creep in automatically, that is difficult to be completely sure of. The sad part is that the stereotypes people have are much easier to identify and correct than the stereotypes a machine or AI algorithm may have.
2. **Lending**: As algorithms become more complex, more opaque and more powerful, the more ways there are for people to be disqualified from loan applications. Someone living in a low-income community, for exasmple, is likely to have friends and family with similar income levels. It's more likely that someone in their extended network would have a poor repayment history than someone in the network of an upper-middle-class white-collar worker—if a scoring algorithm took that fact into account, it might lock out the low-income person just based on his or her social environment.[34]
3. **Crime**: It is not a unique experience that some people are called more frequently for an in-person frisking by security personnel at the airports. But, this can get darker. Through machine learning and algorithms, researchers have repeated the historic criminology experiment of telling criminals apart from law-abiding people using facial recognition.[35] The use of Physiognomy, the ability to judge a person's character from appearance alone, could be dangerous and the people who are supposed to respond quickly can make grave mistakes by relying too much on either instinct or AI.

Stereotypes and discrimination based on them are almost as old as civilization. However, the difference with AI is that these drawbacks may become institutionalized and then may get stronger and stronger. And machines that are expected to be free of them may not even get the blame, or be subject to media brouhaha, or trolling on social media. The problem will then be much larger. In this context, it is important to highlight that there is already some progress being made on these discussions; for example, the Institute of Electrical and Electronics

Engineers (IEEE) has launched the IEEE Global Initiative on Ethics of Autonomous and Intelligent Systems.[36]

The IEEE Initiative on Ethics of Autonomous and Intelligent Systems

The Institute of Electrical and Electronics Engineers (IEEE) is a professional association and was formed in 1963 from the amalgamation of the American Institute of Electrical Engineers and the Institute of Radio Engineers. Today, it is the world's largest association of technical professionals with more than 420,000 members in over 160 countries. Its objectives are the educational and technical advancement of electrical and electronic engineering, telecommunications, computer engineering and allied disciplines.[37]

The IEEE Global Initiative on Ethics of Autonomous and Intelligent Systems is an incubation space for new standards and solutions, certifications and codes of conduct, and consensus building for the ethical implementation of intelligent technologies. The stated objective of this initiative is to ensure every stakeholder involved in the design and development of autonomous and intelligent systems is educated, trained and empowered to prioritize ethical considerations so that these technologies are advanced for the benefit of humanity.

The IEEE Global Initiative on Ethics of Autonomous and Intelligent Systems (A/IS) was launched in April 2016 to move beyond the paranoia and the uncritical admiration regarding autonomous and intelligent technologies and to illustrate that aligning technology development and use with ethical values will help advance innovation while diminishing fear in the process. The goal of the IEEE Global Initiative is to incorporate ethical aspects of human well-being that may not automatically be considered in the current design and manufacture of A/IS technologies and to reframe the notion of success so that human progress can include the intentional prioritization of individual, community and societal ethical values.[38]

The IEEE Global Initiative has two primary outputs—the creation and iteration of a body of work known as *Ethically Aligned Design: A Vision for Prioritizing Human Well-Being with Autonomous and Intelligent Systems*; and the identification and recommendation of ideas for Standards Projects focused on prioritizing ethical considerations in Autonomous and Intelligent Systems (A/IS).

Notes

1 "Albert Einstein: The Human Side". www.nytimes.com/1993/04/18/opinion/l-when-the-twain-of-science-and-god-meet-einstein-s-faith-576093.html (Accessed on April 11, 2018)

2 Stijn Claessens and Laura Kodres, "The regulatory responses to the global financial crisis: Some uncomfortable questions", IMF Working Paper (WP/14/46). www.imf.org/external/pubs/ft/wp/2014/wp1446.pdf (Accessed on November 22, 2017)

3 Charles Owen "Chuck" Prince III is a former chairman and chief executive officer of Citigroup. He succeeded Sandy Weill as the chief executive officer of the firm in 2003, and as the Chairman of the Board in 2006.
4 Michiyo Nakamoto and David Wighton, "Citigroup chief stays bullish on buy-outs", *Financial Times*, July 9, 2007, www.ft.com/content/80e2987a-2e50-11dc-821c-0000779fd2ac?mhq5j=e2 (Accessed on November 22, 2017)
5 "Citi Watch: A king's ransom for prince's exit?", Dealbook, *The New York Times*, Nov. 4, 2007.
6 Steve Denning, "Lest we forget: Why we had a financial crisis", *Forbes*, Nov. 22, 2011, www.forbes.com/sites/stevedenning/2011/11/22/5086/#e620e88f92f1 (Accessed on November 22, 2017)
7 Pat Garofalo, "Mayor Bloomberg: It was not the banks that created the mortgage crisis", *ThinkProgress.org*, https://thinkprogress.org/mayor-bloomberg-it-was-not-the-banks-that-created-the-mortgage-crisis-5eba647c32c7 (Accessed on November 22, 2017)
8 John Carney, "Everything you ever wanted to know about bank leverage rules", *CNBC*, Jul. 12, 2013, www.cnbc.com/id/100880857 (Accessed on November 22, 2017)
9 Antonio Regalado, "Google's AI explosion in one chart", *MIT Technology Review*, March 25, 2017, www.technologyreview.com/s/603984/googles-ai-explosion-in-one-chart/ (Accessed on 22nd November 2017). "The democratization of machine learning: What it means for tech innovation", April 13, 2017, http://knowledge.wharton.upenn.edu/article/democratization-ai-means-tech-innovation/ (Accessed on November 22, 2017)
10 For a discussion on how regulations can inhibit or stimulate technological change, see: Jonathan B. Weiner, "The regulation of technology, and the technology of regulation", *Technology in Society Journal*, doi:10.1016/j.techsoc.2004.01.033 (Accessed on November 22, 2017)
11 Discussion on regulations and emerging technologies, with focus on nanotechnology: Armin Wiek, Stefan Zemp, Michael Siegrist, and Alexander I. Walter, "Sustainable governance of emerging technologies – Critical constellations in the agent network of nanotechnology", *Technology in Society Journal*, doi:10.1016/j.techsoc.2007.08.010 (Accessed on November 22, 2017)
12 IPCC Climate Change. "The physical science basis. Contribution of working group I to the fifth assessment report of the intergovernmental panel on climate change". (2013): 1535. (Accessed on November 22, 2017)
13 www.moneyflowindex.org/indian-solar-sector-tops-the-global-fund-raising-of-3-2-billion-till-now-in-2017/3231722/ (Accessed on November 22, 2017)
14 Paul Berg, David Baltimore, Sydney Brenner, Richard O. Roblin III, and Maxine F. Singer,"Summary statement of the Asilomar Conference on Recombinant DNA Molecules", *Proceedings of the National Academy of Sciences of the United States of America*, 72(6): 1981–1984, June 1975. www.ncbi.nlm.nih.gov/pmc/articles/PMC432675/pdf/pnas00049-0007.pdf (Accessed on November 22, 2017)
15 Brynjolfsson, Erik and Andrew McAfee, 2014. *The Second Machine Age: Work, Progress, and Prosperity in a Time of Brilliant Technologies*. New York, NY: WW Norton & Company.
16 "'All in our time'—A foul and awesome display", *Bulletin of the Atomic Scientists*, 31(5); One point of view can be seen here: Zachary Keck, "Why nuclear weapons work", Sept. 13, 2014, *The Diplomat*, http://thediplomat.com/2014/09/why-nuclear-weapons-work/ (Accessed on November 22, 2017).A recent perspective can be found here: Edward Ifft, "A challenge to nuclear deterrence", Arms Control Today, Arms Control Association, Mar 2017. www.armscontrol.org/act/2017-03/features/challenge-nuclear-deterrence (Accessed on November 22, 2017)
17 Please see note no 13.

18 http://nationalacademies.org/gene-editing/Gene-Edit-Summit/ (Accessed on November 22, 2017)
19 Information about CRISPR-Cas9 can be found here: www.yourgenome.org/facts/what-is-crispr-cas9 (Accessed on November 22, 2017); Ewen Callaway, "UK scientists gain license to edit genes in human embryos", Feb 1, 2016, *Nature*, www.nature.com/news/uk-scientists-gain-licence-to-edit-genes-in-human-embryos-1.19270 (Accessed on 22 November 2017)
20 www.congress.gov/106/plaws/publ224/PLAW-106publ224.pdf (Accessed on November 22, 2017)
21 https://futureoflife.org/bai-2017/ (Accessed on November 22, 2017)
22 Bostrom, Nick. (2005). "Transhumanist values." *Journal of Philosophical* Research, 30, Supplement: 3–14. (Accessed on November 22, 2017)
23 Jerry Useem, "How online shopping makes suckers of us all", *The Atlantic*, May 2017. www.theatlantic.com/magazine/archive/2017/05/how-online-shopping-makes-suckers-of-us-all/521448/ (Accessed on November 22, 2017)
24 In economics, a free market is an idealized system in which the prices for goods and services are determined by the open market and consumers, in which the laws and forces of supply and demand are free from any intervention by a government, price-setting monopoly, or other authority. www.businessdictionary.com/definition/free-market.html (Accessed on April 11, 2018)
25 In microeconomics, supply and demand is an economic model of price determination in a market. It postulates that in a competitive market, the unit price for a particular good, or other traded item such as labor or liquid financial assets, will vary until it settles at a point where the quantity demanded will equal the quantity supplied, resulting in an economic equilibrium. www.investopedia.com/university/economics/economics3.asp (Accessed on April 11, 2018)
26 An AI takeover is a hypothetical scenario in which artificial intelligence (AI) becomes the dominant form of intelligence on Earth, with computers or robots effectively taking control of the planet away from the human species. www.forbes.com/sites/shephyken/2017/12/17/will-ai-take-over-the-world/#610814805401 (Accessed on April 11, 2018), https://futureoflife.org/2015/11/23/the-superintelligence-control-problem/ (Accessed on April 11, 2018)
27 Robert Hart, "Saudi Arabia's robot citizen is eroding human rights. What happens when the rights of humans become devalued?" Feb. 14, 2018. https://qz.com/1205017/saudi-arabias-robot-citizen-is-eroding-human-rights/ (Accessed on February 15, 2018)
28 Harold Stark, "Artificial intelligence and the overwhelming question of human rights", *Forbes*. July 19, 2017. www.forbes.com/sites/haroldstark/2017/07/19/artificial-intelligence-and-the-overwhelming-question-of-human-rights/#626993376c90 (Accessed on February 15, 2018)
29 Christine Wong, "Top Canadian researcher says AI robots deserve human rights", Oct. 11, 2017. "Should artificially intelligent robots have the same rights as people? Yes, says one of Canada's top artificial intelligence (AI) entrepreneurs. Suzanne Gildert is co-founder and chief scientific officer of Kindred AI, a Vancouver startup whose backers include Google's venture capital arm. At the SingularityU conference in Toronto, Gildert made the case for extending human rights to artificially intelligent robots. "A subset of the artificial intelligence developed in the next few decades will be very human-like. I believe these entities should have the same rights as humans," said Gildert." www.itbusiness.ca/news/top-canadian-researcher-says-ai-robots-deserve-human-rights/95730 (Accessed on February 15, 2018).
30 Cate Brown, "The rise of artificial intelligence and the threat to our human rights", July 14, 2017. "From Frankenstein's murderous monster to the malevolent force of the Matrix, science fiction often tussles with the idea of an evil other: a human creation

gone wrong. These creations have so far kept to the world of fiction, with artificial intelligence framed in largely positive terms. Google is the pub quiz master, Siri is on hand to make restaurant reservations and, by 2021, we'll have our own virtual chauffeurs. But where is artificial intelligence heading? Could science fiction become reality? And what does this all mean for human rights?" https://rightsinfo.org/rise-artificial-intelligence-threat-human-rights/ (Accessed on February 15, 2018).

31 https://dictionary.cambridge.org/dictionary/english/stereotype (Accessed on February 15, 2018).

32 www.apa.org/monitor/julaug04/criminal.aspx (Accessed on April 11, 2018).

33 Sarah Griffiths, "It's not just humans, COMPUTERS can be prejudiced too: Software may accidentally sort job applications based on race or gender", *MailOnline*, Aug. 17, 2015. "Lots of firms now use computer software to sort through job applicants. Computer scientists at the University of Utah have devised a test which reveals whether an algorithm could be biased like a human being. If the test can predict a person's race or gender based on hidden data in their CV, there is a potential problem for bias, but it can be fixed. Currently it's a proof of concept but could one day be used by recruiters." www.dailymail.co.uk/sciencetech/article-3200811/It-s-not-just-humans-COMPUTERS-prejudiced-Software-accidentally-sort-job-applications-based-race-gender.html (Accessed on February 15, 2018)

34 Kaveh Waddell, "How algorithms can bring down minorities' credit scores: Analyzing people's social connections may lead to a new way of discriminating against them", *The Atlantic*, Dec. 2, 2016. www.theatlantic.com/technology/archive/2016/12/how-algorithms-can-bring-down-minorities-credit-scores/509333/ (Accessed on February 15, 2018)

35 Katyanna Quach, "AI can now tell if you're a criminal or not", *The Register*, 18 Nov. 2016 www.theregister.co.uk/2016/11/18/ai_can_tell_if_youre_a_criminal/ (Accessed on February 15, 2018)

36 http://standards.ieee.org/develop/indconn/ec/autonomous_systems.html (Accessed on February 15, 2018)

37 www.ieee.org// (Accessed on February 15, 2018)

38 https://standards.ieee.org/develop/indconn/ec/ec_about_us.pdf (Accessed on February 15, 2018)

6

POLICY RESPONSE HAS TO EVOLVE IN PARALLEL

> Everything that civilization has to offer is a product of human intelligence; we cannot predict what we might achieve when this intelligence is magnified by the tools that AI may provide, but the eradication of war, disease, and poverty would be high on anyone's list. Success in creating AI would be the biggest event in human history. Unfortunately, it might also be the last.
>
> —*Stephen Hawking*[1]

The urgent need to prepare for more AI and its impact

We can't say with certainty if AI will ultimately prove to be a friend or an enemy. But, there is no such thing as "over preparation". Never ever in history has over preparing for a true friend's welcome led to regret or has overestimating the enemy been a disadvantage. There is a common saying in defense forces that is one of the basic principles of training: *the more you sweat in peace, the less you bleed in war.* On the other hand, taking a threat lightly and underestimating it could lead to you remaining underprepared, which could have disastrous consequences. The same principle applies to AI and the emerging threat because of it.

There is no one, single threat. In December 2016, experts on AI and advisors to the President of the United States of America released a report on AI.[2] The report highlights the effects of automation on jobs and paints a gloomy picture on risks to existing jobs and a further increase in wage inequality between low-skilled and high-skilled workers. If we take the key messages from this report seriously (and there is no reason why we should not), there is a genuine reason to feel concerned that there are a large number of jobs that could potentially be at risk; society needs to be prepared for this eventuality. The report warns the US government that:

> Responding to the economic effects of AI-driven automation will be a significant policy challenge for the next Administration and its successors. AI has already begun to transform the American workplace, changing the types of jobs available and the skills that workers need to thrive. All Americans should have the opportunity to participate in addressing these challenges, whether as students, workers, managers, or technical leaders, or simply as citizens with a voice in the policy debate.
>
> AI raises many new policy questions, which should be continued topics for discussion and consideration by future Administrations, Congress, the private sector, and the public. Continued engagement among government, industry, technical and policy experts, and the public should play an important role in moving the Nation toward policies that create broadly shared prosperity, unlock the creative potential of American companies and workers, and ensure the Nation's continued leadership in the creation and use of AI.

How big will the AI impact be on jobs and what will this lead to?

There is virtually a consensus that machines will continue to get smarter, and more automation that continues to have a greater AI component going forward will be the norm rather than exception, but the house is extremely divided on the how the impact of AI on jobs will play out. There are different estimates (or we would rather call them guesses) depending on the source, and whom you find more credible is a subjective matter. Some of these estimates are benign and some are extremely pessimistic about the AI threat to jobs.

The range of impact is very wide, beginning with an estimate that less than 5 percent of total jobs are at risk over the next 20 years and extending all the way to about half of all jobs being under serious threat. We can continue to argue about the possible impact but the plain and cold reality is that in the last few years, automation and AI have quietly started taking away jobs even in the areas we may not notice on a daily basis. Hence, it is important to look at some of the data that is available, even for these "not the frontline" industries, in a dispassionate manner.

In a November 2017 article, "A new chart conclusively proves that automation is a serious threat", at *Futurism*,[3] an important fact was highlighted about employment at oil rigs. As the number of rigs declined due to falling oil prices, the number of workers the oil industry employed also went down. But when the number of oil rigs began to rebound, the number of workers employed didn't. For example, the increased use of technology has been able to simplify the repetitive task of connecting drill pipe segments to each other as they're shoved deep into the Earth and the task now requires fewer people to accomplish. It is expected that what once took a crew of 20 to complete will soon take a crew of five. The application of new technologies to oil drilling means that of the 440,000 jobs lost in the global downturn, as many as 220,000 of those jobs may be lost permanently.

In the best-case scenario, workers that are being displaced would be able to come back to the mainstream workforce following training and learning new techniques and hence, by adapting to the new environment. The worst-case scenario is that these jobs are lost forever and no amount of adaptability would help. Nevertheless, it is impossible to escape the fact that AI is most likely to have a significant impact. And here, the worst part is that lower-paying jobs that require low skills are at the highest risk in the near term. By definition, a low-skill job is something that can be easily programmed and quickly assigned to machines even when very advanced AI tools are only in their preliminary stages of development.

It is not just the possibility of job losses due to automation but other associated changes that will take place because of AI because of which we immediately need to train workers for the new environment and need to evolve a consistent and coherent policy for this. The question is not just about job losses alone, though that may be the root cause of other problems and at the same time will be the most important symptom of this challenge. We are yet to fully understand how the impact on jobs will impact society in general. But, there is an immediate need to take the AI threat to jobs seriously and to figure out how this can be handled.

But, not everyone is on the same page

There are some extremely vocal supporters of AI who think we should not worry too much about AI. For example, several of them believe that AI's impact on jobs will be marginal and there will be more new jobs that will be created because of AI. Similarly, they believe that, on balance, AI will be more beneficial than the possible harmful effects it might cause. It is not a surprise that these AI supporters include influential politicians, scientists, successful world leaders, technology entrepreneurs, philosophers and AI experts.

US Treasury Secretary, Steven Mnuchin

US Treasury Secretary Steven Mnuchin[4] thinks exactly opposite on the possible AI impact. In March 2017, Mnuchin surprised everyone when he said that he is not worried about AI displacing US jobs for at least 50 years: "I think that is so far in the future—in terms of artificial intelligence taking over American jobs—I think we're, like, so far away from that," said Mnuchin. "Not even on my radar screen," he added. Effectively, the president's top economic advisor categorically declined to say if he had any worries about robots putting people out of work.[5]

Computer scientist and futurist, Ray Kurzweil

Ray Kurzweil,[6] who is a renowned expert on AI, thinks that in the long run AI will do far more good than harm. He is optimistic about the time when computers will surpass human intelligence and singularity will be achieved. We have already eliminated all jobs several times in human history. In a conversation published in *Fortune* in September 2017, Kurzweil replied to a question on the impact of jobs from AI and other technologies by saying:[7]

> We have already eliminated all jobs several times in human history. How many jobs circa 1900 exist today? If I were a prescient futurist in 1900, I would say, "Okay, 38% of you work on farms; 25% of you work in factories. That's two-thirds of the population. I predict that by the year 2015, that will be 2% on farms and 9% in factories." And everybody would go, "Oh, my God, we're going to be out of work." I would say, "Well, don't worry, for every job we eliminate, we're going to create more jobs at the top of the skill ladder." And people would say, "What new jobs?" And I'd say, "Well, I don't know. We haven't invented them yet."

Peter H. Diamandis, Chairman of the X Prize Foundation and the co-author of Abundance: The Future Is Better Than You Think

Peter Diamandis[8] is one of the most optimistic AI experts. In one of his articles,[9] the influential technologist wrote:

> Artificial Intelligence (AI) is a massive opportunity for humanity, not a threat. AI will level the global playing field. In the future, AI will democratize the ability for everyone to have equal access to services ranging from healthcare to finance advice. And likely it will do all of these things for free, or nearly for free, independent of who you are or where you live. Ultimately, AIs will dematerialize, demonetize and democratize all of these services, dramatically improving the quality of life for 8 billion people, pushing us closer towards a world of abundance.
>
> Why I Don't Fear AI (At Least, Not For Now). First of all, we (humans) consistently overreact to new technologies. Our default, evolutionary response to new things that we don't understand is to fear the worst. Nowadays, the fear is promulgated by a flood of dystopian Hollywood movies and negative news that keeps us in fear of the future. In the 1980s, when DNA restriction enzymes were discovered, making genetic engineering possible, the fear mongers warned the world of devastating killer engineered viruses and mutated life forms. What we got was miracle drugs, and extraordinary increases in food production.
>
> The Benefits Outweigh the Risks. AI will be an incredibly powerful tool that we can use to expand our capabilities and access to resources. In short,

humanity will ultimately collaborate and co-evolve with AI. When we talk about all of the problems we have on Earth, and the need to solve them, it is only through such AI-human collaboration that we will gain the ability to solve our grandest challenges and truly create a world of abundance.

Mark Zuckerberg, co-founder, Chairman and CEO of Facebook

Mark Zuckerberg[10] is highly optimistic about the potential of AI. He thinks that it could benefit humankind in several ways. At the time of his debate on the future impact of AI with other tech entrepreneurs, Zuckerberg wrote[11]:

> One reason I'm so optimistic about AI is that improvements in basic research improve systems across so many different fields—from diagnosing diseases to keep us healthy, to improving self-driving cars to keep us safe, and from showing you better content in News Feed to delivering you more relevant search results. ... Every time we improve our AI methods, all of these systems get better. I'm excited about all the progress here and its potential to make the world better.

He also said, "I think people who are naysayers and try to drum up these doomsday scenarios—I just, I don't understand it. It's really negative and in some ways I actually think it is pretty irresponsible."

Eric Schmidt, Executive Chairman of Alphabet Inc. (parent of Google)

Eric Schmidt[12] was initially skeptical about the technology, and he's since acknowledged how vital it is to both the company's mission and to the global economy.[13] "I was proven completely wrong" about AI, Alphabet's executive has chairman said. "General AI," or mimicking the elasticity of human thought, is still decades away from reality, by Schmidt's estimation. But he has become more bullish about the prospect in recent years. As far as questions about apocalyptic scenarios, like a robot uprising, Schmidt thinks that these are important philosophical questions, but ones that we're not facing right now. In Schmidt's view, the positives far outweigh the negatives for AI. "Things that bedevil us, like traffic accidents and medical diagnoses will get better," he said.

Satya Nadella, Chief Executive Officer at Microsoft

As per one of the articles on *CNBC*, Satya Nadella[14] rebutted claims that AI will accelerate wealth disparity and instead said that it could be a vital driver of growth.[15] Nadella believes that the world needs technological breakthroughs like the emerging AI industry to kickstart weak economic growth. "It's not like we actually have economic growth today. So we actually need technological breakthrough, we need AI," he added. He said that it was a crucial challenge for tech companies to make sure that AI led to inclusive growth. "We should do our very best to train people for the jobs of the future," he said in response to fears that AI could lead to job losses. "Our responsibility is to have the AI augment the human ingenuity and augment the human opportunity. I think that's the opportunity in front of us and that's what we have got to go to work on", he said.

Ray Kurzweil, Peter H. Diamandis, Mark Zuckerberg and Eric Schmidt are only some among many of the thought leaders who think that AI will be good for humankind. These supporters of AI think that very good things can happen in the near future as a result of exponential leaps in technology, and that AI will bring forth a utopian future, in which the human brain's full potential will be opened up and there will be new solutions to all of humanity's problems, including the "eradication of disease and poverty."[16] But is this too benevolent a view on AI?

Are we picking pennies in front of a steamroller?

In economics and finance, a *Taleb distribution* is the statistical profile of an investment that normally provides a payoff of small positive returns, while carrying a small but significant risk of catastrophic losses.[17] The term was coined by journalist Martin Wolf and economist John Kay to describe investments with a "high probability of a modest gain and a low probability of huge losses in any period." The concept is named after Nassim Nicholas Taleb, based on ideas outlined in his books *The Black Swan* and *Fooled by Randomness*. It occurs in a speculative bubble, where one purchases an asset in the expectation that it will likely go up, but may plummet, and hope to sell the asset before the bubble bursts. This is referred to as **"picking up pennies in front of a steamroller"**.

"Picking up pennies in front of a steamroller" means steadily making investments for a small return, only to be eventually crushed by a financial crash or a "steamroller".[18] If there are supporters of AI who think that AI will be good for us, there are perhaps more legends who believe that AI's impact on jobs will be massive and not just that, many of them also believe that AI poses an "existential risk"[19] to humankind. As for gains from AI, they believe that the benefits of AI are similar to humankind picking up pennies in front of a steamroller. ***Why are so many big names from industry and academia worried about AI and who are they?***

Stephen Hawking, renowned physicist and popular author

One of the most popular scientists in contemporary physics, Stephen Hawking,[20] was an English theoretical physicist, cosmologist, author and Director of Research at the Centre for Theoretical Cosmology, University of Cambridge. His scientific works include collaboration with Roger Penrose on gravitational singularity theorems in the framework of general relativity and the theoretical prediction that black holes emit radiation.

Stephen Hawking was of the view that AI needs to be tackled with caution and in a correct way. Otherwise, AI has all the potential to become a danger and could be a huge threat to humankind. From time to time, Hawking talked about the increasing role of robots and machines, and how this may lead to unintended and disastrous consequences for humanity. He thought that AI could be an existential threat to humankind.[21]

In December 2014, BBC News spoke to Hawking[22] and in that interaction, Hawking expressed his concern about the future impact of AI. He thought that the biological evolution of the human race is very slow. So, it would not be surprising that the human race could easily be surpassed by AI. According to him, AI will be "either the best, or the worst thing, to ever happen to humanity". Hawking believed that AI, if used properly, could help us to eradicate diseases and poverty and control the damage to the climate and natural environment. But if we don't control the direction and pace of progress in AI, this may backfire very easily.

This is not incorrect, conceptually speaking. The human brain has developed over millions of years as a result of complex processes of natural selection and evolution. This cannot change over a short span of time, which means that even centuries and thousands of years are not enough to make changes in how the human brain operates. There is absolutely no comparison between humans and AI machines or AI robots in terms of how fast they can evolve.

Elon Musk, entrepreneur and CEO of Tesla and SpaceX

Elon Musk[23] is the CEO of Tesla and SpaceX. Musk, in collaboration with Sam Altman, founded OpenAI, a billion-dollar non-profit company, to work for safer AI. Musk is also one of the high-profile investors along with Mark Zuckerberg, Jeff Bezos, Vinod Khosla and Ashton Kutcher in Vicarious, a company aiming to build a computer that can think like a person, with a neural network capable of replicating the part of the brain that controls vision, body movement and language.

Elon Musk thinks that humans need to merge with machines to remain relevant. In an age when AI is certain to become more prevalent and could make humans useless, there is a need to merge with machines, according to Musk. This may require an implant in the human brain, which can help in accessing and processing information quickly.[24] His view is that over time, we will see a closer merger between biological intelligence and digital intelligence. He thinks that humans will find it difficult to compete with AI machines and that there are several areas where humans simply can't compete with machines. Computer AI machines, for example, can communicate at an unimaginably faster rate compared to humans. Musk's advice in many ways seems similar to "if you can't beat them, join them".

Musk has expressed his concern about "deep AI," which goes beyond specific applications such as better shop-floor operations, efficient language translation or driverless cars. "Deep AI" or "artificial general intelligence" could make possible machines that are "smarter than the smartest human on earth". And that makes AI our biggest existential threat. We need to have regulatory oversight at the national and international level to avoid the AI disaster that is not just possible but very likely if AI remains unchecked. Elon Musk says that AI should be regulated. He advises that we should be very careful about AI because in his view it is the biggest existential threat to humankind. In 2014, Musk's views on the dangers of AI became widely known when he spoke at MIT. Musk has drawn an analogy that in history, there are enough examples where scientists have become so engrossed in their work that they don't really realize the implications of what they're doing.

The existential threat may be a more long-term concern but the immediate worry is job losses. Musk thinks that the immediate threat is how AI will displace jobs. He predicts that as much as 15 percent of the global workforce could become unemployed in the not-too-distant future. According to Musk, the most near-term impact is autonomous cars and he thinks that it is going to happen much faster than what we are prepared for. Musk argues that it is imperative to figure out new roles for a huge number of drivers because driverless cars will make them redundant. This will be very disruptive and happen very quickly, Musk believes.

Bill Gates, Co-Founder of Microsoft and Co-Chairman of Bill & Melinda Gates Foundation

Microsoft founder and one of the wealthiest and most successful businesspeople in the world, Bill Gates,[25] is concerned about the existential threat from super-intelligent machines. Gates thinks that AI is potentially more dangerous than a nuclear catastrophe. He has expressed that we should be very careful about AI when the machines become superintelligent and more powerful. Gates thinks that in the beginning, the machines will do a lot of jobs for us and that will be helpful. However, as the machines progress and become more intelligent, it will become a bigger challenge to handle.[26] However, he has also said that AI advancements will outweigh any potential pitfalls in some of his recent statements.

There are other AI-linked immediate challenges that Bill Gates is worried about. There is a risk of job losses, and he also makes another argument that robots who replace human workers should incur taxes equivalent to that worker's income taxes. Gates argues that these taxes, paid by a robot's owners or makers, would be used to help fund labor force retraining and deployment in other areas.[27] For example, factory workers would be transitioned to health services, education, or other fields where human workers will remain in the workforce. These artificial barriers such as tax on AI would intentionally slow down the speed of automation and would give more time for managing the broader transition. Gates notes that the market alone won't be able to deal with the speed of that transition and much of the potential for putting free labor to better use will be in the public sector.

Gates notes that automation won't be allowed to thrive if the public resists it. In his view, taxation is a better way to handle the public sentiment. If automation doesn't benefit all members of society, it could generate massive opposition and that would be detrimental to the advancement of AI technology. Bill Gates also notes that technology could "accentuate" the gap between the rich and poor. He warns that technology advancements could increase the gulf if technology is only taught in rich schools, and if it's expensive it could not benefit everyone.[28]

Nick Bostrom, philosopher and author

University of Oxford Philosophy professor, Nick Bostrom,[29] is one of the most influential voices on AI. He warned in his 2014 book, *Superintelligence*, that once unfriendly superintelligence exists, it would prevent us from replacing it or changing its preferences. Our fate would be sealed. Bostrom, who is also the director of the University of Oxford's Future of Humanity Institute, take a dim view of the trend in which the role of AI is increasing, warning that AI could quickly turn dark and dispose of humans, generating "economic miracles and technological awesomeness" within places of work, study and leisure, but with nobody there to benefit at the end. Bostrom has used an apt analogy and called the idea of a society of technological awesomeness with no human beings a "Disneyland without children".[30]

Irrespective of the view you hold, there needs to be consideration of the potential impact of AI on jobs and the workforce. AI and automation are relevant for the economy and jobs and there is hardly anyone who can argue against this.

Agent Orange and you may not always be happy with what you invent

It's not unusual for inventors, researchers, and developers to rue their creations. Albert Einstein regretted his role in the development of the atomic bomb. Alfred Nobel, who developed dynamite, launched the Nobel Peace Prize to award promoters of peace. Mikhail Kalashnikov was the inventor of one of the deadliest guns, the AK-47. The guns were cheap, lightweight, and durable in the most extreme climates, earning Kalashnikov hero status in Russia and becoming a source of pride for most of his life.

Unfortunately, when terrorist groups and other violent factions started using the AK-47, Kalashnikov wished he could have made something more helpful. By the time he was near the end of his life, Kalashnikov became so concerned by the bloodshed that he wrote to the head of the Russian Church to beg forgiveness:

> My spiritual pain is unbearable. My heartache unbearable. I keep having the same insoluble question: if my rifle deprive people of life, and therefore I,

> Mikhail Kalashnikov, ninety-three years old, the son of a peasant, and Orthodox Christian according to his faith, [am] responsible for the death of people, even an enemy?[31]

However, there is a different side as well. Arthur Galston's research led to the development of Agent Orange, a herbicide initially intended to help grow crops but that caused extensive environmental damage as well as birth defects when used in Vietnam. He said in 2003, "nothing that you do in science is guaranteed to result in benefits for mankind. Any discovery, I believe, is morally neutral and it can be turned either to constructive ends or destructive ends. That's not the fault of science."[32]

There is another dimension. The path of progress in scientific inventions is rarely linear. In May 2017, a driver was killed in a self-driving Tesla whose sensors failed to notice the tractor-trailer crossing its path. Although an investigation by the National Highway Traffic Safety Administration later found that Tesla's Autopilot system was not to blame, it is not always possible to think of what the path of invention will look like. There is no certainty that we will reach our intended destination and reach it in our desired way; and when we are dealing with something as powerful as AI, it may be just too dangerous.

AI and humans: Relationship is not as simple as "master and servant"

While the *Master Algorithm*[33] remains an exciting possibility in the development of AI, it is important to realize that AI machines do not have to become self-aware and human-like before we take them and their implications seriously. AI just has to become intelligent enough, which it already has, in order for it to impact society in a very serious way. AI already has crossed that threshold of minimum intelligence. It will only become more and more ubiquitous now. And that is what is different this time. We are at a critical juncture with regard to our response regarding how we handle AI—research-wise and regulations-wise—going forward.

There are those who say that since AI is being developed by humans, how can AI machines beat humans in intelligence? It is like saying that since cars were developed by humans, they should not be able to run faster than humans. Or that a robot developed by humans cannot be stronger or more efficient than humans. The IBM Deep Blue was developed by programmers who were themselves not great chess players but they were still able to develop a machine that was able to beat the reigning world champion. For example, since advanced AI uses algorithms that are more complex than the human brain can understand, it could be a threat. This could be true even when the scientists behind this technology are human and are in control of this technology.

There are some people, and this includes AI researchers and scientists, who don't agree with the assertion that AI could ever be a risk to humans. Their view is that the scientists who develop AI will remain in charge of what they develop, thus AI does not seem to pose any type of threat to the human race. Instead of a problem, AI can help us in solving complex problems, a distinct advantage that was never available to humans in their history over millions of years. The supporters also think that it is too premature to worry about AI becoming a threat. Despite continuous and huge advancements in AI, we are nowhere close to achieving AI that is either equal or superior to humans.

There is a need to have a comprehensive policy on AI

So far, public policy on AI is virtually non-existent and the developments do not come under the purview of any single regulation. In the United States, the Federal Aviation Administration oversees drones, the Securities and Exchange Commission oversees automated financial trading, and the Department of Transportation has begun to oversee self-driving cars.[34] Unless there is a consistent and comprehensive policy to deal with these situations, there is a high likelihood that things may go out of control at some point and that could be a *point of no return*.

In case of job losses and the resulting impact on individuals and societies, there is a need to have a plan and policy on how the companies, industries and countries would prepare for it. Since there is so much variance in how AI's impact on jobs would play out, it is understandable that there will be a recurring need to go back to the drawing board. But, that is better than having no plan at all. The dangers of AI are not limited to misuse, or AI in itself becoming uncontrollable. Even in a relatively benign scenario where AI is only substituting workers, the challenge requires a comprehensive response.

Similarly, when AI replaces some jobs, it will also be possible that it will create new tasks and new jobs that may be different than what we have seen so far. The responsibility would be collective to get ready for these new opportunities whenever they arise. If AI machines are going to be everywhere at some point in the future, it is imperative that we need regulation by a representative body that can oversee the development and deployment of AI. AI needs to be regulated to ensure that it remains under the control of humans.

Things can go wrong and can become uncontrollable because AI is progressing too quickly. It is much more acceptable to go a bit slower but in the right direction in comparison to going faster but being uncertain that the direction is right. Traveling faster in the wrong direction is equivalent to running eagerly and faster towards apocalypse. In addition, there is always a chance of a rogue state or non-state actors using AI to damage humanity. The regular occurrence of virus attacks on computer systems globally is proof enough to show how vulnerable we are.

There are already autonomous weapons and hacking algorithms that can be used to make systems go haywire. AI in the wrong hands can lead to a disaster. As we have seen in the case of Facebook's influence in recent elections globally, any uncontrolled technological initiative could become a very powerful tool in the hands of demagogues to propagate populist ideas that could cause long-term harm. Just like Hitler and many more dictators or even democratically elected leaders, the negative impact of AI on jobs will be a useful area of emphasis to mobilize public opinion and support authoritarian rule.

How AI and jobs could be linked: The lessons from industrialization

The improvement in the economic status of the masses over the previous 250-odd years is mostly linked with the spread of industrialization. This helped not just the absolute prosperity levels in the industrialized nations but also took wealth to a greater number of people. The growth in both GDP and GDP per capita is proof of this. There are two other uniquely interesting features of industrialization. One is good and the other is less so.

The first one is that the pace of industrialization in each subsequent phase was higher than in the previous one.[35] Industrialization and its positive impact on wealth creation took root much more quickly in the United States than Europe. Japan and South Korea were faster than even the United States and the fastest and an unprecedented large-scale example of industrialization is China. It was like an artform that was perfected over time and whose inefficiencies were eliminated.

The second feature is that industrialization was not an automatic process. It was not as if it could happen on its own. Some countries benefited much more than others and a large number remained more or less untouched. Some of the oil-rich Asian countries (their wealth could be attributed to a "chance" event more than anything else) and some others in Africa and Latin America remained virtually untouched by industrialization.

The interesting part is that irrespective of the extent of its geographical spread and specific location, industrialization in general benefited almost all the countries that made an effort towards it, and was successful. Essentially, whichever country did industrialization the right way, by focusing on where the export opportunities are[36] and where the unskilled labor can be employed in large quantities by shifting them from low-productivity areas to higher-productivity manufacturing, made greater gains.[37]

These findings have important implications for how AI will impact jobs in society. It would be a grave mistake to assume that AI will automatically lead to greater prosperity, peace and well-being of humankind. Just like industrialization, there will be a right way to move forward with AI by focusing on opportunities and threats in a methodical and diligent manner. This means that there will be a need for policy that will govern AI and its future development. Humankind could well benefit from AI but without a proper policy, it looks less probable.

The protection of information: Need for a safe and secure environment?

In this context, it is important to highlight some of the policy developments that are directly or indirectly linked to different aspects of AI/automation. For example, the General Data Protection Regulation[38] (GDPR) regulation tries to address inaccurate or discriminating automated decisions by giving individuals the right to request a review of an automated decision. GPDR is a regulation by which the European Parliament, the Council of the European Union and the European Commission intend to strengthen and unify data protection for individuals within the European Union (EU).

GPDR also addresses the export of personal data outside the EU, and in essence this regulation aims to give control back to citizens and residents over their personal data and to simplify the regulatory environment for international business. The regulation was adopted on April 27, 2016. It became enforceable from May 25, 2018 after a two-year transition period and, unlike a directive, it does not require national governments to pass any enabling legislation, and is thus directly binding and applicable.

In the case of GPDR, there are several important initiatives to ensure the intended outcome. But, we would still highlight the Data Processing and Consent part of GPDR. In case of Data Processing, there has to be a lawful basis, e.g. the data subject has given consent to the processing for one or more specific purposes, or processing is necessary for the performance of a contract or for the performance of a task carried out in the public interest. The next interesting part is Consent. Where consent is used as the lawful basis for processing, consent must be explicit for data collected and the purposes the data are used for. Consent for children must be given by the child's parent or custodian, and must be verifiable. Data controllers must be able to prove "consent" (opt-in) and consent may be withdrawn. GPDR does not work in isolation and the EU is not the only authority that is worried about this issue. The EU's GDPR will coexist with a multitude of similar laws enacted across the globe.[39] The laws include:

- The Children's Online Privacy Protection Act (United States)
- Personal Information Protection and Electronic Documents Act (Canada)
- The Health Insurance Portability and Accountability Act (United States)
- State Data Breach Laws (United States)
- Privacy Framework and Cross-Border Privacy Rules (APEC)
- Federal Data Protection Law Held by Private Parties and Its Regulations (Mexico)
- Personal Data Protection Act (Singapore)
- Cybersecurity Law (China)
- Data Privacy Act and Its Implementing Rules and Regulations (Philippines)
- Personal Data (Privacy) Ordinance (Hong Kong)

Big changes in big systems: Is AI at the tipping point?[40]

Big and complex systems in most situations display certain "hard to miss" signs of big changes. Also, many a time, the small, incremental changes lead to bigger changes. Bigger changes bring about even bigger and faster changes, and as the rate of change becomes exponential, the entire system changes qualitatively and comprehensively. The development happening in AI is now occurring at an ever faster rate as developments in several fields lead to exponentially fast changes in AI.

In sociology, this is called the *Tipping Point.* The term was first used in Sociology by the University of Chicago political science professor, Morton Grodzins. Motivated by physics, he used this term to describe the sudden dramatic change in the behavior of a group when a previously rarely used practice becomes widespread. Are we at such a juncture, a tipping point, in the usage of AI in our society? The problem with tipping points is that it is like forecasting a stock market crash.

Tipping points are easy to talk about, but the exact timing is very hard to predict. And in the field of AI, a number of things have been happening over the last few years, as discussed in this book, which indicate that we might be close to an irreversible impact. When we say we are at a critical juncture, we especially stress the criticality with regard to policy actions. The consequences of automation and AI on the jobs market is so big, and so irreversible, that we must act now.

We keep coming back to jobs again and again. This is in fact natural. Nothing else leads to mass discontent in society as much as job loss does. We have seen geopolitical unrest sweep across the world. From the Philippines, to the United States; from Brexit in the UK, to the surge in the popularity of Marine Le Pen's far-right party in France, nationalism is gaining a strong foothold across the globe. Job insecurity is a major concern across populations that helps propel a sense of insecurity and nationalism.

Policy inaction and missteps with the future of AI have the potential to intensify grievances. The time to act is now as tomorrow might be just too late.

Notes

1 In the *Independent* on May 1, 2014, leading scientists and thinkers Stephen Hawking, Stuart Russell, Max Tegmark and Frank Wilczek wrote an article titled "Transcendence looks at the implications of artificial intelligence – but are we taking AI seriously enough?". Stephen Hawking was the director of research at the Department of Applied Mathematics and Theoretical Physics at the University of Cambridge and a 2012 Fundamental Physics Prize laureate for his work on quantum gravity. Stuart Russell is a computer-science professor at the University of California, Berkeley and a co-author of *Artificial Intelligence: A Modern Approach.* Max Tegmark is a physics professor at the Massachusetts Institute of Technology (MIT) and the author of *Our Mathematical Universe.* Frank Wilczek is a physics professor at the MIT and a 2004

Nobel laureate for his work on the strong nuclear force. The article noted that "there are no fundamental limits to what can be achieved: there is no physical law precluding particles from being organised in ways that perform even more advanced computations than the arrangements of particles in human brains. One can imagine such technology outsmarting financial markets, out-inventing human researchers, out-manipulating human leaders, and developing weapons we cannot even understand. Whereas the short-term impact of AI depends on who controls it, the long-term impact depends on whether it can be controlled at all". www.independent.co.uk/news/science/stephen-hawking-transcendence-looks-at-the-implications-of-artificial-intelligence-but-are-we-taking-9313474.html (Accessed on November 13, 2017)

2 The report *Artificial Intelligence, Automation, and the Economy* was released in December 2016 as a follow-up on the previous report, *Preparing for the Future of Artificial Intelligence*, which was published in October 2016. The December report investigates the effects of AI-driven automation on the US job market and economy, and outlines recommended policy responses. This report was produced by a team from the Executive Office of the President of United States of America including staff from the Council of Economic Advisers, Domestic Policy Council, National Economic Council, Office of Management and Budget, and Office of Science and Technology Policy. The analysis and recommendations included herein draw on insights learned over the course of the Future of AI Initiative, which was announced in May 2016, and included Federal Government coordination efforts and cross-sector and public outreach on AI matters. www.whitehouse.gov/sites/whitehouse.gov/files/images/EMBARGOED%20AI%20Economy%20Report.pdf (Accessed on November 14, 2017)

3 Scott Santens, "A new chart conclusively proves that automation is a serious threat", *Futurism*, Nov, 5, 2017. The article notes that "We've ignored manufacturing being automated. Yes, we know it happened, but we've pretended that everyone just went on to find new paid work, without critically evaluating the nature of that paid work. Unemployment isn't a problem, right, because the unemployment rate is at a record low? Tell that to the person who went from a 40-hour per week career with benefits and a sense of security to three different jobs/gigs without any benefits, working 80 hours per week to earn less total income in a far more insecure life just trying to get by each month." https://futurism.com/universal-basic-infrastructure/ (Accessed on November 14, 2017)

4 Steven Terner Mnuchin is the United States Secretary of the Treasury as part of the Trump administration. After his graduation from Yale University in 1985, Mnuchin worked for famous investment bank Goldman Sachs. After he left Goldman Sachs in 2002, he worked for and founded several hedge funds. Mnuchin joined Trump's presidential campaign in 2016, and in February 2017 was confirmed to be Secretary of the Treasury. He has been a vocal supporter of proposed tax reforms and is also an advocate for cuts in corporate tax rates in the United States of America.

5 During a conversation with Mike Allen of *Axios*, US Treasury Secretary Steven Mnuchin said that there was no need to worry about AI taking over jobs for next 50–100 years. www.businessinsider.in/Treasury-Secretary-Mnuchin-says-job-stealing-AI-is-so-far-in-the-future-that-its-not-even-on-my-radar-screen-heres-why-hes-wrong/articleshow/57831903.cms (Accessed on November 15, 2017), www.cnbc.com/2017/03/24/mnuchin-treasury-secretary-ai-automation-comments.html (Accessed on November 15, 2017), www.theatlantic.com/business/archive/2017/03/mnuchin-ai/520791/ (Accessed on November 15, 2017)

6 Raymond "Ray" Kurzweil is an American author, computer scientist, inventor and futurist. Kurzweil is also involved in fields such as optical character recognition (OCR), text-to-speech synthesis, speech recognition technology, and electronic keyboard instruments. He has written books on AI, technological singularity, and futurism. Kurzweil has an optimistic outlook on life extension technologies and the future of nanotechnology, robotics, and biotechnology.

7 Michal Lev-Ram, "Why futurist Ray Kurzweil isn't worried about technology stealing your job", *Fortune*, September 24, 2017. In this conversation, Kurzweil also said that technology has always been a double-edged sword and although technology amplifies both our creative and destructive impulses, it is very clear that overall human life has gotten better. He also said that all of these technologies are a risk, including the most powerful and impactful ones such as biotechnology, nanotechnology, and AI but if we look at history, we're being helped more than we're being hurt. http://fortune.com/2017/09/24/futurist-ray-kurzweil-job-automation-loss/ (Accessed on November 15, 2017)
8 Peter H. Diamandis is a leading and well-known name in innovation, technology and commercial space. He earned an undergraduate degree in Molecular Genetics and a graduate degree in Aerospace Engineering from MIT, and received his M.D. from Harvard Medical School. In 2014, he was named one of "The World's 50 Greatest Leaders" by *Fortune* magazine. He is a Greek American engineer, physician and entrepreneur best known for being the founder and chairman of the X Prize Foundation, the co-founder and executive chairman of Singularity University, and the co-author of bestsellers *Abundance: The Future Is Better Than You Think* and *BOLD: How to Go Big, Create Wealth, and Impact the World*. He is also the former CEO and co-founder of the Zero Gravity Corporation, the co-founder of Space Adventures Ltd., the founder and chairman of the Rocket Racing League, the co-founder of the International Space University, the co-founder of Planetary Resources, and co-founder of Human Longevity, Inc. www.diamandis.com/about (Accessed on November 15, 2017)
9 Peter Diamandis regularly writes on technological and contemporary issues and he is a popular commentator on these subjects. www.diamandis.com/blog/ai-bring-it-on (Accessed on November 15, 2017)
10 Mark Elliot Zuckerberg is an American computer programmer and internet entrepreneur. He is a co-founder of Facebook, and currently operates as its Chairman and Chief Executive Officer.
11 Mark Zuckerberg's comments on AI were widely covered by media during his public spat with other technology entrepreneurs on the future potential and threats as a result of new technological advancements in AI. www.cnbc.com/2017/07/26/mark-zuckerberg-defends-a-i-again-continuing-debate-with-elon-musk.html; https://www.cnbc.com/2017/07/24/mark-zuckerberg-elon-musks-doomsday-ai-predictions-are-irresponsible.html, www.recode.net/2017/7/25/16026184/mark-zuckerberg-artificial-intelligence-elon-musk-ai-argument-twitter; http://fortune.com/2017/07/26/mark-zuckerberg-argues-against-elon-musks-view-of-artificial-intelligence-again/ (Accessed on November 15, 2017)
12 Eric Emerson Schmidt is an American software engineer, a businessman, and the Executive Chairman of Alphabet Inc. From 1997 to 2001, he was Chief Executive Officer (CEO) of Novell. From 2001 to 2011, Schmidt served as the CEO of Google. He has served on various other boards in academia and industry, including the Board of Trustees for Carnegie Mellon University, Pennsylvania, and Princeton University.
13 Robert Hackett, "Alphabet's Eric Schmidt: 'I was proven completely wrong' about artificial intelligence", *Fortune*, Feb. 16, 2017. http://fortune.com/2017/02/15/eric-schmidt-rsa-artificial-intelligence/ (Accessed on November 15, 2017)
14 Satya Narayana Nadella is an Indian American business executive. He is the current Chief Executive Officer (CEO) of Microsoft, succeeding Steve Ballmer in 2014. Before becoming CEO, he was Executive Vice President of Microsoft's cloud and enterprise group.
15 Matt Clinch, "Microsoft CEO Nadella: We have no global growth, we need AI", Jan. 17, 2017, *CNBC*. www.cnbc.com/2017/01/17/microsoft-ceo-nadella-we-have-no-global-growth-we-need-ai.html (Accessed on November 15, 2017)
16 Dominic Basulto, "Why the world's most intelligent people shouldn't be so afraid of artificial intelligence", *Washington Post*, Jan. 20, 2015. This article talks about the other side

of "existential risk" which is "existential reward"—the possibility that AI could significantly help in improving quality of life. www.washingtonpost.com/news/innovations/wp/2015/01/20/why-the-worlds-most-intelligent-people-shouldnt-be-so-afraid-of-artificial-intelligence/?utm_term=.c7190abcfd49 (Accessed on November 15, 2017)

17 https://drive.google.com/file/d/0B8nhAlflk3QIR1o1dnk5ZmRaaGs/view (Accessed on November 15, 2017)

18 www.barrypopik.com/index.php/new_york_city/entry/picking_up_nickels_in_front_of_a_steamroller (Accessed on November 15, 2017)

19 An existential risk is a risk posing permanent large negative consequences to humanity that can never be undone. In Nick Bostrom's paper on the subject, he defined an existential risk as: *One where an adverse outcome would either annihilate Earth-originating intelligent life or permanently and drastically curtail its potential.* The total negative impact of an existential risk is one of the greatest negative impacts known. Such an event could not only annihilate life as we value it from Earth, but it would also severely damage all Earth-originating intelligent life potential.

20 Stephen Hawking was the former Lucasian Professor of Mathematics at the University of Cambridge and author of *A Brief History of Time*, which was an international best-seller. He was then the Dennis Stanton Avery and Sally Tsui Wong-Avery Director of Research at the Department of Applied Mathematics and Theoretical Physics and Founder of the Centre for Theoretical Cosmology at Cambridge. In 1963, Hawking contracted motor neurone disease and was given two years to live. Yet he went on to Cambridge to become a brilliant researcher. He was a fellow of the Royal Society and a member of the US National Academy of Science. Stephen Hawking is regarded as one of the most brilliant theoretical physicists in contemporary Science. www.hawking.org.uk/ (Accessed on November 15, 2017)

21 Stephen Hawking's views on the dangers of AI. www.independent.co.uk/life-style/gadgets-and-tech/news/stephen-hawking-artificial-intelligence-fears-ai-will-replace-humans-virus-life-a8034341.html, www.independent.co.uk/life-style/gadgets-and-tech/news/stephen-hawking-artificial-intelligence-could-wipe-out-humanity-when-it-gets-too-clever-as-humans-a6686496.html, www.independent.co.uk/life-style/gadgets-and-tech/news/stephen-hawking-artificial-intelligence-could-wipe-out-humanity-when-it-gets-too-clever-as-humans-a6686496.html, www.huffingtonpost.in/entry/stephen-hawking-ai-artificial-intelligence-dangers_n_6255338 (Accessed on November 15, 2017)

22 Rory Cellan-Jones, "Stephen Hawking warns artificial intelligence could end mankind", *BBC News*, Dec. 2, 2014.. www.bbc.com/news/technology-30290540 (Accessed on November 15, 2017)

23 Elon Musk is a South African-born Canadian-American businessman, investor, engineer and inventor. Musk is known for his ground-breaking, unconventional, courageous ideas. The Tesla and SpaceX CEO, Musk is passionate about interplanetary colonization and launches reusable rockets into space, hoping to eventually inhabit the Red Planet. He has announced plans to send space tourists on a flight around the moon shortly. He dreams of a world powered by cheap and abundant solar energy. He is the founder, CEO, and CTO of SpaceX; co-founder, CEO, and product architect of Tesla Inc.; co-founder and chairman of SolarCity; co-chairman of OpenAI; co-founder of Zip2; and founder of X.com, which merged with Confinity and took the name PayPal. He was among the Top 100 wealthiest people in the world in December 2016, ranked on the Forbes list of "The World's Most Powerful People". He is passionate about reducing global warming and reducing the "risk of human extinction" by "making life multi planetary" by setting up a human colony on Mars.

24 Maureen Dowd, "Elon Musk's billion-dollar crusade to stop the A.I", *Vanity Fair*, April 2017. www.vanityfair.com/news/2017/03/elon-musk-billion-dollar-crusade-to-stop-ai-space-x, Ruth Umoh, "Why Elon Musk might be right about his artificial

intelligence warnings", *CNBC*, Aug. 25, 2017. www.cnbc.com/2017/08/25/why-elon-musk-might-be-right-about-his-artificial-intelligence-warnings.html, Steven Finlay, "We should be as scared of artificial intelligence as Elon Musk is", *Fortune*, Aug. 18, 2017. http://fortune.com/2017/08/18/elon-musk-artificial-intelligence-risk/ (Accessed on November 15, 2017)

25 William Henry Gates III, more popularly known as Bill Gates, is an American businessman and philanthropist. One of the richest persons in the world, Bill Gates is the co-founder of Microsoft, one of the world's largest technology companies. Through his foundation, he along with his wife, Melinda, has been very active in doing philanthropic work. Though he has remained associated with Microsoft as an owner, he has delegated the executive responsibilities to professional managers.

26 Peter Holley, "Bill Gates on dangers of artificial intelligence: 'I don't understand why some people are not concerned'", *Washington Post*, Jan. 29, 2015. www.washingtonpost.com/news/the-switch/wp/2015/01/28/bill-gates-on-dangers-of-artificial-intelligence-dont-understand-why-some-people-are-not-concerned/ (Accessed on November 15, 2017)

27 Kevin J. Delaney, "The robot that takes your job should pay taxes, says Bill Gates", *Quartz*, Feb. 17, 2017. https://qz.com/911968/bill-gates-the-robot-that-takes-your-job-should-pay-taxes/, David Z. Morris, "Bill Gates says robots should be taxed like workers", *Fortune*, Feb. 18, 2017. http://fortune.com/2017/02/18/bill-gates-robot-taxes-automation/ (Accessed on November 15, 2017)

28 Hadley Gamble and Matt Clinch, "Bill Gates says technology could 'accentuate' the gap between the rich and poor", *CNBC*, Nov. 15, 2017. www.cnbc.com/2017/11/14/bill-gates-defends-the-rise-of-the-robots.html (Accessed on November 15, 2017)

29 Nick Bostrom is a Swedish philosopher known for suggesting that future advances in AI research may pose a supreme danger to humanity, if the problem of control has not been solved before superintelligence is brought into being. Bostrom cautions that even when given an innocuous task, a superintelligence might ruthlessly optimize, and destroy humankind as a side effect. He says that although there are potentially great benefits to be obtained from AI, the problem of control should be the absolute priority.

30 Fabrizio Cardinali, "A Disneyland without Children?" *LACE Project*, Aug. 22, 2016. "In this post, LACE workplace learning technologies expert, Fabrizio Cardinali, takes a look at the revamped interest on Artificial Intelligence, and how new trends such as Machine Learning, Robotic Process Automation and Cognitive Technologies are emerging. which might bring together a greater world to work and learn within, but perhaps with far less (human) workers than smart machines around. A great Disneyland for all of us...but perhaps a Disneyland without Children left to amuse." www.laceproject.eu/blog/disneyland-without-children/ (Accessed on November 15, 2017)

31 http://bigthink.com/laurie-vazquez/6-scientists-who-regret-their-greatest-inventions (Accessed on November 15, 2017)

32 Arthur W. Galston, Agent Orange researcher, passed away in 2008. Galston, a Yale plant biologist who did early research that helped lead to the herbicide Agent Orange, helped raise awareness of the military's use of it in Vietnam in the 1960s and its devastating effects on river ecosystems. www.nytimes.com/2008/06/23/us/23galston.html (Accessed on November 15, 2017)

33 *The Master Algorithm: How the Quest for the Ultimate Learning Machine Will Remake Our World* is a book by Pedro Domingos released in 2015. Towards the end of the book, he pictures a "master algorithm" allowing technology to allow machine-learning algorithms to asymptotically grow to a perfect understanding of how the world and people in it work.

34 Please see note no. 24, "Elon Musk's billion-dollar crusade to stop the A.I".

35 Mary Hallward-Driemeier and Gaurav Nayyar. 2017. *Trouble in the Making? The Future of Manufacturing-Led Development*. Overview booklet. Washington, DC: World Bank. License: Creative Commons Attribution CC BY 3.0 IGO.

36 Pierre-Richard Agénor and Otaviano Canuto. 2015. "Middle-income growth traps." *Research in Economics*, 69(4): 641–60.
37 Margaret S McMillan and Dani Rodrik. 2011. *Globalization, Structural Change and Productivity Growth.* Working Paper No. 17143. Cambridge, MA: National Bureau of Economic Research (NBER).
38 After four years of preparation and debate, GDPR was finally approved by the EU Parliament on April 14, 2016. Enforcement date: May 25, 2018—at which time those organizations in non-compliance may face heavy fines. The EU General Data Protection Regulation (GDPR) replaces the Data Protection Directive 95/46/EC and was designed to harmonize data privacy laws across Europe, to protect and empower all EU citizens' data privacy and to reshape the way organizations across the region approach data privacy. www.eugdpr.org/ (Accessed on February 14, 2018), www.itgovernance.co.uk/data-protection-dpa-and-eu-data-protection-regulation (Accessed on February 14, 2018), https://ico.org.uk/for-organisations/guide-to-the-general-data-protection-regulation-gdpr/ (Accessed on February 14, 2018)
39 BRINK editorial staff, "The great matchup: EU's data law vs. the world", *BRINK*, Aug. 4, 2017. www.brinknews.com/the-great-matchup-eus-data-law-vs-the-world/ (Accessed on February 14, 2018)
40 http://whatis.techtarget.com/definition/tipping-point (Accessed on February 22, 2018), www.merriam-webster.com/dictionary/tipping%20point (Accessed on February 22, 2018)

7

THE ASILOMAR PRINCIPLES FOR ARTIFICIAL INTELLIGENCE

> We must address, individually and collectively, moral and ethical issues raised by cutting-edge research in artificial intelligence and biotechnology, which will enable significant life extension, designer babies, and memory extraction.
>
> —*Klaus Schwab*[1]

AI and the need for guiding principles

The technologies that could disrupt social and economic harmony and cause massive job losses, and as some naysayers believe, could even put the very survival of humanity under threat, certainly need some basic guiding principles. No one would want to put a complete stop or reverse the cycle on these technologies because they are useful and if used in a better way could help humankind, but it is difficult to argue that they don't require oversight. Whether anything new will help or harm us depends less on the absolute characteristics of that technology and more on how it is being used.

This was the basic theme of the previous chapter of this book, Chapter 6, *Policy response has to evolve in parallel.* We have guiding principles and some basic "dos and don'ts" for everything: from as basic as banking and insurance, traffic safety, stopping substance abuse, urban planning, maintaining air and water quality, safe use of medicines, prevention of communicable diseases and control of pollution caused by chemicals and hazardous industrial waste, to as advanced as pharmaceutical research, application of nuclear technology, aviation (civil or military) and even space research. The basic idea is never to put a blanket ban but to move with caution.

The core philosophy is that we have to be extra careful when we are venturing into unchartered territories because "the unknown" is exciting but it could be dangerous when we are not paying attention.. As long as we are doing something that we don't understand well and that is evolving rapidly and could even have a remote possibility of affecting others, guiding principles are needed. Drunk driving is a crime not because you could harm yourself while driving drunk; the far bigger reason is that there is a possibility of you harming others on the road while driving drunk. Similarly, there is hardly any debate concerning the fact that AI will impact others and not just on the people developing and working on these technologies.

It is important to clarify here that we are not ruling out that this may be a false alarm. Naturally, we can't deny with absolute confidence that it is possible that the future impact of these new technologies in areas like jobs, humankind's survival, and their usage as unfair and unethical tools in shaping public opinion may have been exaggerated. It may also be true that the impact of new AI technologies will be higher in IT/IT enabled services (ITES), financial services, transportation, e-commerce and retail in comparison to some areas in manufacturing such as precision engineering, textiles and others where human intervention is necessary. But, there is no case for ignoring the possibilities of what AI could do in the worst-case scenario.

The doomsday scenarios may not always play out but when they do, no preparation is adequate. Ultimately, AI is like any other tool but much more powerful and much less understood. History has enough examples to tell us that law breakers are more often a step ahead of law enforcers, and more powerful and less understood tools have a very high probability of being misused and abused. We agree that this can happen with any technology and the risk of abuse can be used as a bogey by both sides: those who oppose and those who defend; for good and for evil, for genuine purpose and for selfish reasons. But, there will be a need for guiding principles irrespective of the case.

Asilomar: Will it do to AI what it did to DNA?

Asilomar is not a very old place but that does not mean that it does not have a rich history and that some of the discussions that took place at this "not so well-known location" will not have a defining impact on the areas in which these deliberations took place and in the world in general. Asilomar State Beach, with its 107 acres of beach and conference grounds area, is a three-hour drive from the hustle and bustle of San Francisco, the hub of the modern internet economy. The word "asilomar" is a combination of two Spanish words: "asilo", meaning asylum, and "mar", meaning the sea. The state park is also known as Monterey Peninsula's "Refuge by the Sea".

Its history dates back to 1913 when it served as the grounds for the Young Women's Christian Association (YWCA) leadership camp. There is another historic landmark in the region, the famous Hearst Castle. Built by the architect Julia Morgan, the castle served as a residence for the media magnate William

Randolph Hearst. Hearst's mother, Phoebe Hearst, was on the committee of the Pacific Coast YWCA. Due to this connection, Julia Morgan was also hired by the YWCA to design the conference grounds in Asilomar for its events and conference.[2]

With its origins in YWCA's progressive ideas, perhaps it is only fitting that over time the conference grounds were used to host a whole series of conferences with further cutting-edge ideas. A number of workshops and conferences took place in Asilomar over the years including the 1975 Asilomar Conference on Recombinant DNA, the 2010 Asilomar international conference on climate intervention technologies, and more recently, the 2017 *Beneficial AI* conference.

The 1975 recombinant DNA conference is perhaps the most well known of all the Asilomar conferences. The organizers of the conference took seriously not just the scientific, but also the ethical, implications of genetics and biotechnology, and displayed great foresight by opening up to the public the discussion regarding the immediate and long-term impacts of biotechnology on society and the planet.[3] The conference offered a good template on which many of the later discussions in the field were based.

The 1975 recombinant DNA conference was highly influential not only for industry and academia, but also the general public. The guidelines developed by the discussions helped scientists conduct research and experiments over the subsequent years. The conference also helped increase public debate about biomedical research and genetics, fostering knowledgeable discussion about the social, political and environmental issues emerging from biotechnology. The field avoided the kind of rhetoric that afflicts some of the other more recent discussions, such as climate change.

Asilomar Principles for AI

The *Future of Life Institute* is a Boston-based think tank founded by Skype co-founder Jaan Tallinn and physicist Anthony Aguirre. The think tank focuses on the social impact of technologies and ways to mitigate potential negative outcomes. The institute particularly looks at the existential risk coming from AI, biotechnology, nuclear technology, and climate change. It has an impressive list of supporters and donors, many of whom are well-known personalities, such as Elon Musk, Alan Alda and Morgan Freeman. It also has academicians like Nick Bostrom and Erik Brynjolfsson among others, and previously Stephen Hawking, on its scientific advisory board.[4]

The selection of location may have been symbolic but the idea was serious.The Future of Life Institute organized a three-day conference in January 2017 at Asilomar, the same location where the Asilomar Conference on Recombinant DNA had taken place in 1975, to discuss ways to help guide future AI development efforts in a direction that is beneficial to mankind. The conference was called *Beneficial AI*. The conference concluded with the formation of a number of defining principles aimed at helping AI develop along positive lines, and to help research progress and unfold in a way that keeps humans in control. The principles were divided into three distinct categories: research, ethics and values, and longer-term issues.[5]

The conference participants came from an extremely wide spectrum of professions and had different and varied backgrounds. Developing a set of principles that all the attendees could agree on was but an easy process. The principles were developed through a process of idea generation, ironing out disagreements and contradictions through multiple discussions. As a starting point, the organizers used a number of recent AI reports from academia, government, industry and the non-profit sector to prepare a list of opinions about what society should do to manage AI. This list contained plenty of ambiguities and contradictions.[6]

The meeting attendees were sent this list prior to the actual meeting, and were asked for inputs, evaluations and suggestions for making a more coherent list of principles. Finally, at the meeting, the principles were further improved and finalized through a two-stage process. In the first stage, smaller sets of principles were assigned to subgroups of participants for further refinement. And in the second stage, the entire set of principles was presented to all the participants for their criticism. After this process, a set of principles remained with a high level of support from the attendees.

This list of 23 principles was then put forth as the AI Asilomar Principles. In the rest of this chapter, we go over these principles and discuss their significance. The principles are organized under three categories: (i) research related, (ii) related to ethics and values, and (iii) principles concerned with futuristic/long-term issues.

- Research related – Principle 1 to 5
- On ethics and values – Principle 6 to 18
- On futuristic and long-term issues – Principle 19 to 23

Asilomar Principles on AI research

The first five principles deal with research. This set of principles addresses the research and developmental aspect of AI. We are certain that readers may not fail to notice that these principles may seem straightforward and necessary; hence, what is so special about them? But, the key pint here is that the implementation details can be complicated.

1 Research Goal: The goal of AI research should be to create not undirected intelligence, but beneficial intelligence.
2 Research Funding: Investments in AI should be accompanied by funding for research on ensuring its beneficial use including critical questions in computer science, economics, law, ethics, and social studies, such as:

> How can we make future AI systems highly robust, so that they do what we want without malfunctioning or getting hacked?
>
> How can we grow our prosperity through automation while maintaining people's resources and purpose?
>
> How can we update our legal systems to be more fair and efficient, to keep pace with AI, and to manage the risks associated with AI?
>
> What set of values should AI be aligned with, and what legal and ethical status should it have?

3 Science-Policy Link: There should be constructive and healthy exchange between AI researchers and policy makers.
4 Research Culture: A culture of cooperation, trust, and transparency should be fostered among researchers and developers of AI.
5 Race Avoidance: Teams developing AI systems should actively cooperate to avoid corner-cutting on safety standards.

It seems obvious that there ought to be safeguards in AI against hacking. But what all this would entail is less clear. Would it require the kind of effort and vigilance that the current computer systems require, or would it be something more complex? As AI develops, hackers will probably start developing their own AI systems, hacker AI, to attack beneficial AI. How can one monitor and be vigilant against advanced hacker AI systems? Much more difficult, in our opinion, is the question of growing prosperity through automation, and at the same time maintaining people's resources and purpose. This will require technologists and policy makers to come together and preemptively make plans. For example, instead of being reactive to job loss, we need to be forward-looking and plan introducing AI in a way that adheres to this basic framework.

High-frequency trading

We cannot emphasize enough that the previous track record of policy keeping up with technological developments is not very promising. Scientific and technological developments can happen much faster than the ability of regulatory bodies and policy makers to keep track of these changes. A glaring example, in the field of finance, is high-frequency trading. High-frequency trading is proof enough of how difficult it can be for policy makers to act in a timely manner.

This superfast way of trading, with automation and AI, spread at Wall Street at an amazing speed. The adoption of frequency trading was much faster than what the regulation was prepared for. Regulatory bodies had a difficult time keeping up with the latest advances in this corner of trading. What also made things difficult was a lack of consensus among practitioners, many of whom contradicted each other. There can be powerful vested interests at play, making the field very opaque for outsiders.

High-frequency trading also highlights the problems that can arise if a culture of cooperation, trust and transparency does not exist. This problem is especially sinister in the highly secretive world of finance and investment. There is an understandable reason for the secrecy to safeguard trade secrets. But both for the good of the financial system, as well as for the benefit of the investor, there needs to be transparency and trust.

Legal and ethical issues

Legal and ethical issues for AI can quickly lead to uncharted territory. Consider the fourth item under the second principle. What about the moral and ethical status of AI? In fact, before discussing AI, what are the human characteristics that are considered necessary for moral status? Is there any chance of robots having these characteristics at some point in the future? In their insightful paper, "The ethics of artificial intelligence", Nick Bostrom and Eliezer Yudkowsky have conducted an analysis of some of these issues.[7]

For humans, two relevant characteristics that endow moral status are Sentience and Sapience. Sentience is the capacity for individual, conscious experience. Living beings feel joy, pain, or suffering due to this quality. This capacity is by no means confined to only humans, of course. Sapience, on the other hand, refers to a higher and evolved intelligence. Humans display sapience. Humans are self-aware, can do logical reasoning, and many other activities that are not directly related to the ability to simply survive.

Based on these considerations, current-day AI does not have either Sentience or Sapience, and hence it does not have moral status. But what about future forms of AI that might have one or both of these characteristics? What about AI that has self-awareness, but cannot feel pain or sorrow? What about AI that can also feel pain or sorrow? Should such an AI be considered to have a moral status and harming it as equivalent to harming a human being? What if AI and humans start competing for the affection of the same human or same AI?

Asilomar Principles on AI ethics and values

This second set of principles goes further than the first set, which focused on research-related issues, and addresses questions relevant to the interface between AI and human society. The social impact of a new technology can be a complex issue to analyze. In AI, the discussion is made even more difficult due to the numerous possible future directions in which AI may evolve. The Asilomar Principles from No. 6 to No. 18 deal with AI's ethics and values.

6 Safety: AI systems should be safe and secure throughout their operational lifetime, and verifiably so where applicable and feasible.
7 Failure Transparency: If an AI system causes harm, it should be possible to ascertain why.
8 Judicial Transparency: Any involvement by an autonomous system in judicial decision-making should provide a satisfactory explanation auditable by a competent human authority.
9 Responsibility: Designers and builders of advanced AI systems are stakeholders in the moral implications of their use, misuse, and actions, with a responsibility and opportunity to shape those implications.
10 Value Alignment: Highly autonomous AI systems should be designed so that their goals and behaviors can be assured to align with human values throughout their operation.
11 Human Values: AI systems should be designed and operated so as to be compatible with ideals of human dignity, rights, freedoms, and cultural diversity.
12 Personal Privacy: People should have the right to access, manage and control the data they generate, given AI systems' power to analyze and utilize that data.
13 Liberty and Privacy: The application of AI to personal data must not unreasonably curtail people's real or perceived liberty.
14 Shared Benefit: AI technologies should benefit and empower as many people as possible.
15 Shared Prosperity: The economic prosperity created by AI should be shared broadly, to benefit all of humanity.
16 Human Control: Humans should choose how and whether to delegate decisions to AI systems, to accomplish human-chosen objectives.
17 Non-subversion: The power conferred by control of highly advanced AI systems should respect and improve, rather than subvert, the social and civic processes on which the health of society depends.
18 AI Arms Race: An arms race in lethal autonomous weapons should be avoided.

Some of the issues mentioned in this set of principles have already become extremely relevant. Principles about Safety, Responsibility, Personal Privacy, and Liberty clearly address important issues that are already being discussed to some extent. Safety is clearly one of the most important criteria that must be addressed before introducing AI or automation to any workplace or home environment. It is also important for the designer of an AI to effectively and transparently troubleshoot when AI does fail the safety standards.

As machine learning and AI tools become more complex, this can become a fairly difficult task. Given a set of input data, AI will process it to yield some output data. Humans tend to process the same input data differently, and in many instances arrive at very different conclusions. Humans also tend to display leaps of intuition, gut feeling and guesswork that are not logical methods of deriving conclusions from given input data. As AI is designed to be more human-like, will there be similar possibilities? Principles regarding failure transparency and judicial transparency relate to these issues.

Humans are the chief creators of AI right now. Theoretically, there is also the possibility of AI designing and creating other forms of AI. Who is to be held responsible for the actions of any form of AI? To what extent is an autonomous AI fully independent and fully responsible? If an instance of AI ends up being harmful to humans, who is to be held responsible? Clearly the designer and creator of AI should share a sense of responsibility for the impact on humans and society.

Some of the principles in this second category seem clear enough. But some of the other principles will require further detailed discussion. Implementing them might even be a big challenge. Think about Principles No. 14 and #15 about Shared Benefit and Shared Prosperity. In a free market system guided by capitalism, how can we effectively adhere to these principles? Similarly, discussions about Value Alignment and Human Values, Principles No. 10 and 11, will take on a political dimension depending on the particular country or regime that discusses it.

Transparency in judicial decision-making, mentioned in Principle No. 8, is of utmost importance. Transparency also needs to be part of more mundane compliance enforcing autonomous systems. As an example, think of the enforcement of regulatory compliance in the banking system. There have been reports recently about bank compliance staff being replaced by autonomous systems.[8] In the aftermath of the 2008 financial crisis, the number of regulatory issues that banks have to navigate, the number of regulations that traders have to comply with, and the best practice guidelines regarding how brokers engage with clients, have grown in volume and complexity.

There are strict guidelines, for example, about trade order flow from clients to brokers. These guidelines are in place to prevent, among other things, future occurrences of price rigging and market manipulation. As autonomous systems have started taking over from compliance staff to monitor these details, who would be finally responsible for any financial mishaps? Who would take the blame? Will it be the autonomous system or the human auditor who might be responsible for doing periodic audits?

Consider an AI being designed to take over certain social functions in the judicial system. Bostrom and Yudkowsky point out how some aspects of the judicial system might not be entirely obvious to the technologist designing the AI.[9] While presiding over any new case, a judge is bound to follow past precedence whenever possible. This makes the judicial system predictable to those that follow it. A lack of this predictability would be detrimental to society. However, a technologist sees science and technology as always evolving and improving. Scientists highly value paradigm shifts where entirely new ways of approaching problems and solving them pave the way for further progress. Adhering to past paradigms and solutions is not typically a technologist's modus operandi.[10]

When talking about ethics, there are some especially complex problems that must be addressed. While it is clear that AI ethics must align with human ethics, one must also recognize that the ethical system has evolved over time. The ethical system of the Ancient Greeks was quite different from the current thinking about human ethics and values. And what is considered normal and ethical now will not necessarily be so in another hundred years. From this, it is natural to conclude that as AI develops and evolves, its ethical system will also evolve. What will determine the evolution of machine ethics? Will there be any random component to this evolution? If an AI is based on algorithms with a random component, then it is possible that its ethical system might display random shifts.[11]

Asilomar Principles related to longer-term issues of AI

One can only speculate where AI will be in 20, 50, or a hundred years. There might be another AI winter in its research and development, like the two we saw before, and it might put pause on progress. Or, perhaps, with Big Data and cloud computing, progress will come fast and furious. In this scenario, we might see tremendous development in the field before hitting another hiatus. But, it can be said with a high degree of confidence that the next few years will bring huge change in the way society interfaces and interacts with AI, and while this may open up new opportunities, it will also bring new challenges and difficult choices.

The final set of AI Asilomar Principles addresses futuristic possibilities. Depending on who you talk to right now, you will hear different estimates for what AI can eventually do and how far it can develop. Some commentators dismiss many of the AI-related claims as belonging to the realm of fiction rather than the factual. Others make bold claims about what AI will be able to do soon. There is no consensus. Nevertheless, it is critically important to consider the longer-term possibilities and impact and the last set of given principles deals with them.

19 Capability Caution: There being no consensus, we should avoid strong assumptions regarding upper limits on future AI capabilities.
20 Importance: Advanced AI could represent a profound change in the history of life on Earth, and should be planned for and managed with commensurate care and resources.
21 Risks: Risks posed by AI systems, especially catastrophic or existential risks, must be subject to planning and mitigation efforts commensurate with their expected impact.

22 Recursive Self-Improvement: AI systems designed to recursively self-improve or self-replicate in a manner that could lead to rapidly increasing quality or quantity must be subject to strict safety and control measures.
23 Common Good: Super-intelligence should only be developed in the service of widely shared ethical ideals, and for the benefit of all humanity rather than one state or organization.

The first principle in the final category suggests that we refrain from making any strong assumptions regarding the ability of future forms of AI. Seeing in the rear view all the advances that happened over the last two decades in this field, and just speculating on the technological possibilities in the near future, makes one realize the futility of making any assumptions about the trajectory of development of this field. Will AI become significant enough to seriously affect the future development of human beings?

To what extent will humans interact, interface, and even combine and fuse with AI? Many tech firms are already researching the possibilities of combining AI capabilities with humans. Humans with AI-enhanced limbs or eyes or even brains? Again, what really is possible and what is impossible? What will become reality and what will always remain a figment of imagination? With these imaginations come the possibilities of an AI machine that is not specifically built for a particular task but that can handle multiple complex tasks.

Artificial General Intelligence

The possibilities of Artificial General Intelligence (AGI) and superintelligence belong to that realm of what is not currently possible, but might become reality some day. AGI refers to that future stage of AI when it starts functionally resembling humans. Humans have the ability to apply intelligent decision-making to multiple tasks. A human child learns to walk, read, eat, talk and eventually masters all the tasks that make us human. AI, meanwhile, can usually do one very specific task very well. Researchers can design AI to do more than a single task.

But we are still far from any form of AI that can approach intelligent decision-making in a more general way. If and when AGI is finally achieved, AI will become human-like. The principles in the final category will then become very relevant. And one might even have to develop further sub-principles to address new issues stemming from the introduction of AGI. But before that, what will AGI be like? Is there a way to assess if an AI machine can be called one that has AGI. How can we ever determine if a particular instance of AI is really AGI? What would we require the AI to do to pass the test for AGI?

There are a number of ideas. The original one is the famous Turing Test. This test suggests that a blindfolded human tester talk to the AI and another human. The tester should try to evaluate which of the two is the machine. If the human tester fails to consistently identify the machine, then the machine can be considered an AGI. Another test, attributed to Steve Wozniak, is called the Coffee Test. According to this test, the machine can be an AGI if it can go to an average American home and make a cup of coffee on its own. Another test, attributed to Ben Goertzel, is the Robot Student College Test, which suggests the AI enroll in a college, take the same courses taken by other human students, and obtain a degree.

The reader can make her own test to determine what would pass for an AGI. But one thing is clear: an AGI will be able to do tasks that we expect an average human being to be able to do. While AGI might currently appear to be a distant possibility, the idea of superintelligence is even more sci-fi-like. To quote the Oxford philosopher, Nick Bostrom, superintelligence is "an intellect that is much smarter than the best human brains in practically every field, including scientific creativity, general wisdom and social skills."[12]

Asilomar Principles: Good starting point but there is a long way to go

When will it happen, we don't know, but it is possible that some day AI might be able to recursively self-improve. As AI software gets better, it would be able to reprogram and improve itself. The improved software will be able to improve itself even better. This would lead to recursive self-improvement, and if this happens at a sufficiently fast pace, it could lead to an intelligence explosion. This asymptotic state of intelligence would be the superintelligent state. The last two Asilomar principles address superintelligence.

Instead of envisaging a future where AI or humans, or a combination of the two, embark on a pursuit of superintelligence unshackled by any constraints, these principles suggest that the pursuit be subject to stringent safety and control measures. The principles also suggest that superintelligence be only developed not just as a scientific and technological curiosity, but in a consciously planned way, and in the pursuit of shared ideals and for the benefit of all of humanity.

It is wise to err on the side of caution. These final few Asilomar principles do precisely that. Experts debate about how intelligent and self-sufficient AI can be. Can AI ever evolve into AGI? Will singularity, the epoch when AI becomes more intelligent than humans, ever be reached? While focusing on the long-term issues, the principles stress the need for continual discussion on where AI is heading and its impact on the future of humanity.

Asilomar AI Principles are a very good starting point, but much more needs to be done. The principles cover a broad set of issues ranging from research, to society, to futuristic possibilities. They provide a good launch pad—where sufficient agreement has been obtained among a fairly heterogeneous set of experts—for detailed research, discussion and commentary regarding the practicalities and implementation of the principles. The path from principle to practice can further cement or fracture the principles.

Notes

1 Professor Klaus Martin Schwab is a German engineer and economist and is best known as the founder and executive chairman of the World Economic Forum (WEF). He founded the European Management Forum in 1971, which in 1987 became the WEF. The WEF is a Swiss non-profit foundation, based in Cologny, Geneva. Its mission is cited as "committed to improving the state of the world by engaging business, political, academic, and other leaders of society to shape global, regional, and industry agendas". In 1971, Schwab published *Moderne Unternehmensführung im Maschinenbau* (Modern Enterprise Management in Mechanical Engineering). He argued that modern enterprise must serve not just shareholders but all stakeholders. He has championed the multi-stakeholder concept since WEF's inception. In 1998, with his wife Hilde, he created the Schwab Foundation for Social Entrepreneurship. www.forbes.com/sites/bernardmarr/2017/07/25/28-best-quotes-about-artificial-intelligence/#6584da0c4a6f, www.weforum.org/about/klaus-schwab, (Accessed on November 12, 2017)

2 https://npgallery.nps.gov/pdfhost/docs/NHLS/Text/87000823.pdf (Accessed on November 22, 2017)

3 Paul Berg, David Baltimore, Sydney Brenner, Richard O. Roblin III, and Maxine F. Singer, 1975. "Summary statement of the Asilomar Conference on Recombinant DNA Molecules". *Proceedings of the National Academy of Sciences of the United States of America*, 72(6): 1981–84. https://authors.library.caltech.edu/11971/1/BERpnas75.pdf (Accessed on June 24, 2018)

4 As per the details available on its website, *Future of Life* is currently focusing on keeping AI beneficial and is also exploring ways of reducing risks from nuclear weapons and biotechnology. FLI is based in Boston. https://futureoflife.org/ (Accessed on November 22, 2017)

5 https://futureoflife.org/bai-2017/ (Accessed on November 22, 2017)

6 https://futureoflife.org/2017/01/17/principled-ai-discussion-asilomar/ (Accessed on November 22, 2017)

7 Nick Bostrom and Eliezer Yudkowsky, 2014. "The ethics of artificial intelligence". In *The Cambridge Handbook of Artificial Intelligence*, ed. by Keith Frankish and William M. Ramsey. Cambridge: Cambridge University Press, pp. 316–34.

8 Martin Arnold, "Banks' AI plans threaten thousands of jobs", *Financial Times*, Jan. 25, 2017. www.ft.com/content/3da058a0-e268-11e6-8405-9e5580d6e5fb (Accessed on November 22, 2017); Kim S. Nash, "Deutsche Bank deploys artificial intelligence to help meet demands of regulatory compliance", *The Wall Street Journal*, Apr. 18, 2017. https://blogs.wsj.com/cio/2017/04/18/deutsche-bank-deploys-artificial-intelligence-to-help-meet-demands-of-regulatory-compliance/ (Accessed on November 22, 2017); Ryan Browne, "UK regulator looking to use A.I., machine-learning to enforce financial compliance", *CNBC*, Jul. 13, 2017. www.cnbc.com/2017/07/13/uk-regulator-looking-to-use-a-i-machine-learning-to-enforce-financial-compliance.html (Accessed on November 22, 2017)

9 Please see note 7.

10 Please see note 7.

11 Please see note 6.

12 Nick Bostrom, "How long before superintelligence", *International Journal of Futures Studies*, 1998. https://philpapers.org/rec/BOSHLB(Accessed on November 22, 2017)

8

DISCUSSION ON AI NEEDS TO BECOME MAINSTREAM

> Forget artificial intelligence – in the brave new world of big data, it is artificial idiocy we should be looking out for.
>
> —*Tom Chatfield*[1]

The elephant in the room

It may be possible to argue in favor or against, but when it began globalization was supposed to be a good thing. Policy makers, corporations and technology giants embraced globalization as a mostly benevolent process, which entailed some pain in the short term for some people, but eventually would lead to big gains for all. This was the basic premise of globalization, a premise that was seldom questioned either by experts or by common people. Indeed, the Western world largely welcomed globalization as a positive change, a change for the greater good.

Doubts trickled here and there. A handful of people wondered about the ways in which globalization interacted with the rest of the economy, asking if there could be unforeseen consequences. But for a while such ponderings remained confined to technical journals. A 2014 Federal Reserve Bank of Dallas publication had a catchy title, *Globalization: The Elephant in the Room That Is No More,*[2] but really just addressed the possibility of a connection between globalization and the challenges in crafting effective monetary policy.

The entire world was struggling with making monetary policy effective at fighting the post-2008-crisis global growth slump. But was globalization an important factor that needed to be taken into account? Such studies were sparse. Even academicians and researchers had not fully investigated the interactions between global trade, monetary policy, and growth. As Federal Reserve Bank of

Dallas President at the time, Richard Fisher noted in 2006: "The literature on globalization is large. The literature on monetary policy is vast. But literature examining the combination of the two is surprisingly small."[3]

Soon, however, a number of geopolitical events transpired in quick succession and drew such debates out of the confines of academia, and into the rough and tumble of electoral politics. Brexit,[4] the US presidential election of 2016,[5] the French election of 2017,[6] and a number of other geopolitical events found globalization squarely at the center of the active public debate. One wonders if the disruptive promises of the recent developments in automation, machine learning and AI need to draw lessons from this experience of globalization.

In a democracy, the issues taken up by politicians are not always determined by how critical they are and how important they will be in the future. More often, politicians take up issues on the basis of electoral potential and whether these issues can strike a chord with the public. For a politician, more usually than not, the determining factors are: whether people will easily identify with the issue, whether the issue will be able to capture the imagination of the general population, and whether there will be an emotional reaction from people that will influence the voters in choosing their electoral preferences.

The Brexit vote: Unexpected result or a retaliation post suffering?

Brexit was one of the most important global events of the year 2016. Around the middle of the year, voters across the United Kingdom participated in a poll to decide if the country should remain in the European Union (EU). The vote was about Britain's potential exit from the EU and was commonly referred to as the "Brexit vote". The voters had to make a simple choice between a Yes and a No. The people of Britain were asked to decide if they wanted Britain to remain within the EU or should Britain leave the EU.

On one side, there were people who were of the view that trade with the EU and the free movement of goods and labor was for the benefit of Britain, and that it was in the country's interest to continue with this arrangement. This group was convinced that the economic integration with the EU had made Britain's companies more competitive, and that a common market for goods and labor was benefiting Britain. There might have been a few negatives as a result of Britain's partnership with the EU, the thinking went, but it was a good thing on balance and there was no need to change the status quo. This group of people wanted "No Brexit", and they wanted Britain to remain as a good and responsible member of the EU.

The other side of the argument saw too many disadvantages, and hardly any benefits. Britain would be better off, they thought, outside the EU. In general, people belonging to this group espoused the view that free trade and free movement of goods and labor was harmful to Britain's economy. And that it was in the country's best interest to break free and discontinue the country's membership of the EU. This group was convinced that economic integration with the Eurozone had made Britain's working class worse off, and that a common market for goods and labor was harming the common man.

People belonging to this group thought that even though there were a few positive results from the integration with the Eurozone, it was negative on balance and there was a need to alter this status quo. One of their main grievances was the labor supply from various parts of Europe and the consequent weakening of the competitive positioning of the local workforce. In their view, this needed to be corrected. This group wanted "Brexit", i.e. for Britain to break free of the EU and to exit the longstanding arrangement of open borders and ease of trade.

Prior to the actual vote, there was continuous and extensive media coverage of the Brexit vote and the possible outcome scenarios. Though there were differences of a few percentage points here and there depending on which way the wind was blowing, broadly speaking, there was a consensus against Brexit. Even though some of these opinion polls and voter surveys were indicating a tough fight between "for Brexit" and "anti-Brexit" factions and an uncomfortably close margin of victory, the majority of pollsters predicted the vote to come out against Brexit. Britain shall continue to remain within the EU, said the pollsters. It might not have been a straightforward and foregone conclusion, but no-Brexit was still considered the most likely outcome.

It was also interesting that most of the intelligentsia, many important politicians cutting across party lines, thinkers, intellectuals and celebrities were canvassing aggressively for an "anti-Brexit" vote and were warning people of the disaster that would result if Brexit happened. These supporters of "No Brexit" held the view that people ought to be rational and should take a long-term view of things. That they ought to support market integration, free trade and unrestricted movement of goods, services and labor, for the benefit of all.

And what was beneficial for the world would also be beneficial for the people of Britain. The idea, therefore, was that there might be some losers or winners on an individual basis, but the sum total of Britain being a part of the EU was still positive. There was also the logic that even if the EU common market was somewhat of a pain in the short term, it was fundamentally a good thing for everyone in the long run, and that things would ultimately take care of themselves on a medium- to long-term basis.

We all know how people eventually voted and how it ended. How it led to a major political upheaval in Britain, including the resignation of the then Prime Minister, David Cameron. The surprising part is not that the people voted for "Brexit" but the fact that the media and everyone concerned failed to read the popular sentiment accurately. With the benefit of hindsight, it is not difficult to see that most people didn't have the inkling that Britain would vote for Brexit.

Both the authors of this book actively monitor geopolitical developments, and their impact on financial markets, on a daily, if not minute-by-minute, basis. A number of financial indicators such as asset volatilities, trade flows, risk reversals, turbulence indices, etc., are supposed to at least partially capture signs of impending geopolitical upheavals. Currency markets, for example, operate 24 hours a day, and usually catch signs of upcoming events well in advance. However, financial participants mostly misread the sentiment on the ground.

In fact, it would be appropriate to highlight some of the interactions one of the authors had, during one of his many research trips to London, with market participants in Britain before the Brexit vote. These were people whose roles require them to have a dispassionate view of the outcome of geopolitical events. The response on what could be the likely outcome of Brexit vote was overwhelmingly in favor of "No Brexit".

But the fallacy of these interactions and what the media was reading on the ground, and in turn reporting, was most likely that of selective hearing. Perhaps the said author only spoke to those people who were willing to talk and were in favor of "No Brexit". Financial markets were overwhelmingly positioned for "No Brexit". Media and pollsters were somehow ignoring the silent majority that was in favor of "Brexit" and was not that actively participating in opinion polls and voter surveys. Still, this same set of people went out and voted for "Brexit" on the day of the polls.[7]

Brexit was a result of a strong public reaction against economic stagnation and lack of new opportunities from a section of population. The voters from this section in Britain who suffered because there were more people competing for the same the number of jobs probably identified labor supply from continental Europe as reason for their suffering.

Election of Donald Trump: The ordinary American strikes back

If Brexit was the most shocking event of the summer of 2016, the year was not done with its political surprises yet and delivered another big one in November. After an extremely fraught election campaign that led to a deep, vertical divide in the country, Donald Trump's election as the President of the United States of America was not just unexpected.[8] This election result also demolished many convenient notions and conventional logic about the public acceptability of political candidates.

The conventional political wisdom on who would prove to be good presidential candidate was based on the premise that: a) the candidate may or may not be able to win the support of different sections of society but should not do anything that antagonizes any particular group; b) the candidate should be friendly towards the media and should remain politically correct; c) the candidate should not go against the norms when talking about foreign policy, about who is an enemy of the United States and who is a friend. However, Trump ran a very unconventional campaign and since it has been a successful effort, his campaign has actually demolished many established principles on how to run a campaign, the commonly accepted and followed "dos and don'ts" and how to succeed.

There are various explanations that can be put forth as the reason for Trump's victory such as: a) the opposition candidate, Hillary Clinton, was not able to challenge Trump effectively and struggled to maintain the support of the core Democrat supporters; b) voters were tired of the status quo politics of the so-called elites and they opted for a unconventional candidate; c) the vote for Donald Trump in reality was not for him; it was against the established order of the day and in favor of change and, hence, Donald Trump as being the beneficiary of this sentiment was completely incidental; d) the traditional capitalist model and market economy driven society approach has failed and this led to a change in people's political preferences, and; e) the perceived benefits of globalization have been cornered by a tiny minority and the majority population is now worse off post-globalization compared to the pre-globalization phase.

The post-event analysis would continue and there would be different interpretations put forward by different analysts, but the fact remains that people voted for Donald Trump and he got elected as the President of the most powerful nation on Earth. People have chosen him over other candidates, and he is the President of the United States of America. The common factor in Donald Trump's election and the Brexit vote was the role of the media and how it got its reading of popular choice wrong. Just like the "Brexit" poll, the surprising part is not that the people voted for Trump as the President of the United States, but the startling fact that the media once again failed to read which way the wind was blowing. In fact, except for only a few polls, the verdict was almost unanimous that Hillary Clinton was winning though the margin of victory kept fluctuating until the actual date of polls.

The general response on what would be the likely outcome of the vote was overwhelmingly against Trump. But again, it was a case of selective hearing and not paying enough attention to what the ordinary voters in the United States wanted. Perhaps the media was again only speaking to those people who were willing to talk and were vehemently protesting against Trump.

The reversal (if any) of globalization is because of adverse public opinion

Donald Trump's election was a result of many contributing factors. However, the influence that the growing public opinion against globalization and the resulting jobs erosion had on the election result cannot be denied. It is difficult to escape the assessment and conclusion that the benefits of globalization accrued selectively and the benefits (if at all there are any) to the people who belong to relatively lower economic strata were simply not commensurate with the benefits derived from globalization by large corporations and people from the top echelons of society.

In the marketing parlance, Trump was effectively able to tap this segment and meet the demand for a political leadership that would at least not see globalization as a one-way street and as an irreversible process. The fact was that there were people who were not happy with the consequences of globalization and the perceived lack of appreciation among the top-level political leadership for the problems this phenomenon has created among some sections of ociety. The interesting thing to note here is that the way globalization and its impact reached common people from academic journals and research circles was proof that with the right presentation and outreach techniques, any topic can become an important election issue.

It can be argued that productivity gains and the benefits of globalization have largely been accrued to the privileged class, especially in developed countries.[9] This is not just linked to the rise in sentiment against globalization but also impacts, more perniciously, overall demand creation in society. As purchasing capacity doesn't get distributed equally, the growth in demand for super-expensive luxury goods increases at a faster rate than the demand for basic goods. This is bad for the modern economy as it is different from all other previous models because it is based on mass distribution of demand, purchasing power and, finally, consumption. This premise comes under threat with rising inequality.

It does appear as if the clock has started to turn back on one of the most powerful and influential concepts that drove the world over the last 70 years, "globalization". We will not delve too much into the public perception part, which is well reflected in the support for Brexit and Donald Trump, and we have discussed it already. But, it is not perception or sentiment alone. Fortunately or unfortunately, the scientific inventions and evolution of technology do not care too much for popular opinion. An important corollary of all this is that these changes are so powerful that we could not stop them even if we wanted to.

Similar to the evolution of globalization and the changes in the thought process on it, there are many developments that we think will be responsible for ongoing and further upcoming changes. They include an ineffective transmission of policies into the real economy and reduced potency of tools available to central bankers and politicians to address the problems and challenges in the economy; waves of nationalism sweeping across the globe; and a preference for the "postponement of pain" at the level of decision makers at almost each and every level.

At this point, we would also like to highlight the drastic reduction in the need for the labor to produce unit output and the role of advancements in robotics and AI in it. It is as simple as the fact that when the most appropriate robots are less expensive than humans, the replacement cycle begins and there is hardly anyone who would argue against the fact that the pace of evolution is infinitely faster in robots compared to that of humans. So, the replacement is irreversible and "labor arbitrage" is under serious threat.

The new and fast-emerging robotics, AI and other labor-saving techniques are strongly linked to globalization. If labor can be replaced with newer technologies like these, the rationale for cheaper labor and labor-cost arbitrage actually goes to a toss. A robot or AI-based algorithm will cost the same at a cheaper labor location and developed countries such as the United States, which has more expensive labor costs. In that case, why would the corporations need their factories in China or Bangladesh or Vietnam, why would tech companies need engineers from India, or why would cheaper labor from Romania and Poland be needed in the UK?

Jean-Claude Juncker, the elephant chart, and why change is not good for everyone

Jean-Claude Juncker[10] is a well-known figure on the global political stage. He was the Prime Minister of Luxembourg from 1995 to 2013. He was also the minister-in-charge for the Finance Ministry. He has been a key figure in matters concerning the European Union, and has served a couple of terms as the President of the European Council. Part of the reason for his popularity is his inclination towards speaking candid statements. Our personal favorite is what he said in 2007. At the time of the euro crisis, Juncker famously said "We heads of governments all know what to do, we just don't know how to get re-elected when we do it".[11] In our view, there is nothing else that captures the current political situation across the world and the popular sentiment better than this.

What has been happening in many countries in Europe, including Greece, for a few years now, what has happened in Britain on Brexit and in the US presidential elections, all of it has one common thread. When prosperity bypasses a few key and reasonably well-off sections of society, there are political repercussions. There is a famous story involving Gautam Buddha[12] and Ananda, one of the more famous disciples of Buddha. The story itself is long, but the idea is that even the best of the sermons and the surest of the paths to salvation are of no use for a hungry man. No one can be generous enough to think about the entire world's welfare when his/her own house is on fire. Recently this has meant more experiments with available political choices through democratic means (and thankfully so, as no one would really want to see France of 1789[13]and Russia of 1917[14] again). A couple of data points first.

The rising gap between productivity and compensation

As per one of the widely noticed reports from the Economic Policy Institute, a non-profit and non-partisan think tank, wages have not really kept pace with productivity increases in the United States,[15] which is the largest economy in the world and a good example to test this hypothesis. The correlation between productivity improvement and wage increment was near perfect for almost 30 years after World War II ended; the wages and benefits of the average worker rose in direct proportion to how much the worker produced per hour. However, the relationship did not remain meaningful for very long.

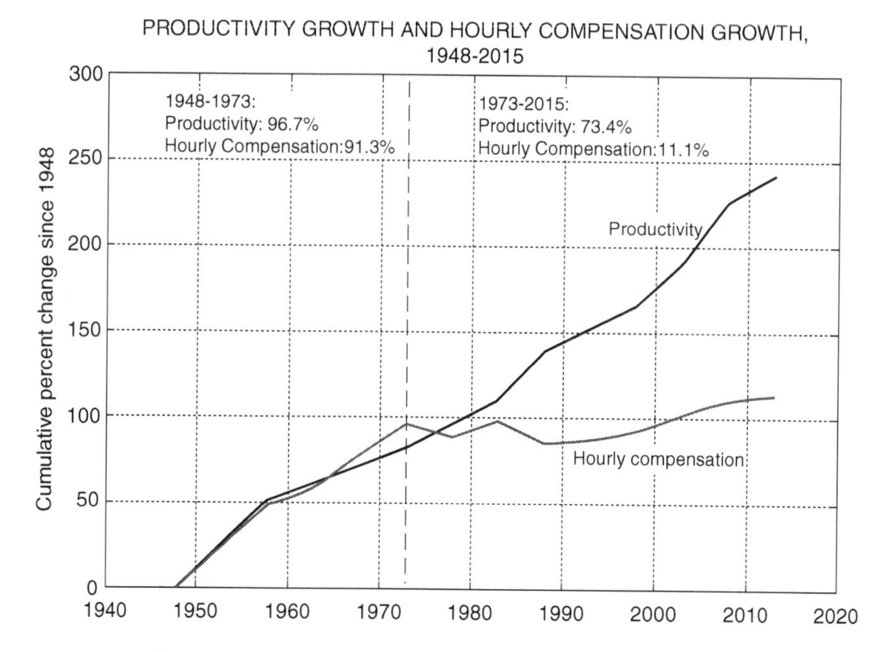

FIGURE 8.1 The relationship between productivity and worker compensation in the United States

Source: Based on *The Productivity-Pay Gap*, Economic Policy Institute

Over the last 40 years or so, while the economic output per hour grew more than 70 percent, the average adjusted-for-inflation wage increase has been a little less than 10 percent. The productivity gains simply didn't come from "human capital". Productivity and wages not only determine the usual cut-and-dry economic fundamentals of a country; they also determine the more messy political evolution. The report goes on to analyze the causes of this breakdown of the linear relationship between productivity and wages, stating that "Rising productivity provides the potential for substantial growth in the pay of the vast majority. However, this potential has been squandered in recent decades". Economic progress has failed to reach a majority of workers, largely because policy choices have favored those pockets of society where income, wealth and power are concentrated. The result of this poor policy craft has been a rise in inequality, the EPI report highlights.

Globalization has not benefited everyone even in developed countries

The second aspect to be discussed is "the elephant chart".[16] This chart, which received a fair share of attention, first appeared in a 2012 World Bank working paper by the economist Branko Milanovic. In his analysis, Milanovic looked at a large number—about 196—of household surveys across the globe. He ranked the world's population, from the poorest 10 percent to the richest 1 percent, in 1988 and again in 2008. At each rank, the chart showed the growth in income between the two years. Looking at the income increases from 1988 to 2008 for various segments of the global population led to a surprising conclusion.

While the consumers in advanced economies benefited from globalization, the average middle class in relatively well-off countries suffered, relatively speaking. This chart, which looks like an elephant, explains how the net result of globalization was negative for the majority of the middle-class population in importing countries, while the benefits went to owners of capital in advanced economies and workers in exporting countries such as China.

One should point out that there are subtleties in making conclusions from the chart. Imagine a country with exactly 100 citizens, with 20 of them making $100 per hour, and 80 of them making the Massachusetts minimum wage, $11 per hour. Imagine also that the income stays flat over the time period. But the population making the minimum wage grows from 80 to 180, while the population making $100 per hour—just the 20 people—stays flat. As we rank people according to their real income, the top 20 percent has now moved to the top 10 percent over the time period. And as they have moved, their original spot in the 10 percent to 20 percent rank interval is now taken up by some of the minimum

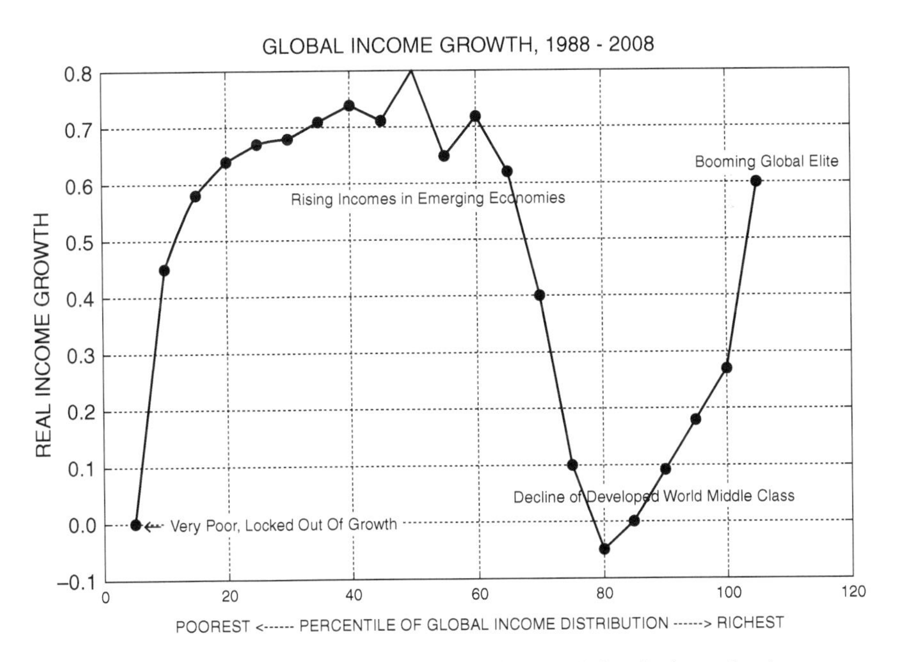

FIGURE 8.2 The distribution of benefits of globalization (The elephant chart)
Source: 2012 World Bank working paper by the economist Branko Milanovic

wage population. Focusing just on this 10%–20 percent part of the chart might then create an illusion of sharply worsening wages for people inhabiting this rank interval of the chart.

However, it is not the same people. It is a new set of people who now inhabit the interval. Milanovic was careful enough to adjust for this population effect between 1988 and 2008. The resulting "fixed" elephant chart wasn't as full of contrasts. However, the message was still the same. There was a segment of middle-class people whose real wages had fallen over this period of heightened globalization. Both these studies point to the difficulty of predicting unintended consequences of policies. Free trade, labor movement, globalization, these are ideas that were studied in detail, debated endlessly, and embraced gingerly. But there is now a clear resistance towards at least some of the ingredients of globalization.

AI has not been part of the mainstream discussions yet

Donald Trump in his election campaign told American workers that he would bring back their jobs by clamping down on trade, off-shoring and immigration. Remarkably, he didn't talk about automation and its impact on jobs. Other presidential candidates didn't talk about automation either. It is understandable why. Technological progress is not always visible to all and is much more difficult to blame. *The Economist* ran a special report recently on the impact of automation and AI on society, and used a story to put into perspective the current particular epoch of AI development, and its public perception.[17]

A long time ago, a man invented the game of chess, and presented it to the king. The king loved it and, to reward the inventor, offered him anything he asked. The savvy inventor asked for one grain of rice for the first square on the chessboard, two for the second, four for the third, and then twice the previous number of grains for each additional square on the chessboard. This adds up to a huge number, about 30 million trillion (3 followed by 19 zeros) grains of rice, which we reckon to be about 1000 trillion pounds of rice, more rice than the annual rice production of the entire world.

The king admitted defeat. The moral of the story is that exponential growth can be unnoticeable, until it suddenly becomes unmanageable. Computing, information technology, and robotics have gone through exponential growth. But what square of the chess board is AI currently at? Are we close to the point when suddenly the politicians start taking notice and the issues stand the risk of becoming politicized?

Certain fears in the minds of the electorate can have a direct bearing on the political fortunes of candidates. Going back to the 2016 US election, did the exposure of the US labor markets to import competition from China have an effect on the outcome? To answer this question in a measurable way, the MIT economist David Autor, and his collaborators, looked for any relationship between the Republican two-party vote share and the growth of Chinese import penetration in the United States over the period 2002 to 2014.[18]

The authors also conducted a counterfactual exercise where they asked how the composition of votes in the closely contested states would have differed had Chinese import penetration grown less than it actually did. The result of this simple study was unequivocal. There is a direct relationship between the Republican vote share and Chinese import growth. The authors noted in their concluding remarks:

> This note relates the change in the county-level Republican two-party vote share between 2000 and 2016 presidential elections to the growth in Chinese import penetration. We find that rising import competition was a robust positive contributor to Republican vote gains. A counterfactual exercise indicates that the Democrat candidate would have won the states of Michigan, Wisconsin, and Pennsylvania—resulting in a majority of votes in the Electoral College—if the growth of import competition from China had only been half as large as actually observed.

This effect on voter preference has not been entirely swift and sudden, manifesting only during the 2016 election. The same authors conducted a previous study where they found a clear contribution of the China shock to the shifting ideological composition of the House of Representatives in the decade leading up to the presidential election. It is entirely possible, as the authors noted, that those representatives' legislative and campaign activities subsequently contributed to the general election outcome in 2016.

One important question to ask is, if the China shock was more psychological in nature, did it just make the electorate wary due to its wide visibility on the counters of Walmart and Target stores across the US? Or was there a real economic impact from the increased competition from Chinese imports? Daron Acemoglu and David Autor of MIT, along with their collaborators, published a study in 2015 that tried to answer precisely this question. They found globalization to be a big factor behind the "US employment sag of the 2000s". According to the study, US-China trade links during the 2000s led to job losses in the range of 2.0–2.4 million. Further, this also led to a weak overall US job growth atmosphere via input-output linkages and other general equilibrium channels.[19]

The point we would like to make is that big changes, economic or technological, have consequences that impact people socially and politically. Globalization and free trade are based on sound principles from an economic point of view. But the political ramifications of these policies are only now beginning to be fully appreciated. Automation and AI are well on their way to bringing about big and disruptive changes. Unintended consequences will definitely occur. But informed discussion and sound policy creation will mitigate unfavorable impact on the people in the middle of the "AI elephant curve". Policy makers are gradually beginning to take note of the challenges brought by a future full of AI and robots. A recent White House study under the Obama administration looked at the impact of AI on employment. It noted the significant present and potential future impact, and observed the need for taking sound policy steps now to mitigate the impact on those who lose their employment to automation. Across the countries, if the recent political fallout of globalization is anything to go by, it is not too early to worry about AI's impact.

Is it time for AI politicians?

Imagine a time in the future when AI becomes an integral part of politics. AI will help guide decision makers by giving them guidance in the form of not just solid facts and figures, but also by suggesting the most likely course of events in the face of uncertainty, the most effective policy path ahead in the absence of sufficient information, and even singling out lawmakers who might be in need of improvement in their performance. After all, algorithms excel over humans in decision-making in the midst of uncertainty. Wouldn't AI make the ideal policy maker? Algorithms have already successfully challenged human hedge fund managers; perhaps algorithms will also start challenging human politicians? Star investment managers, who used to successfully navigate unpredictable markets, used to be humans—Paul Tudor Jones,[20] George Soros,[21] Bruce Kovner,[22] to name a few.

Now the new superstars are the algorithms managing money at firms like Renaissance Technologies, Two Sigma, D. E. Shaw, and other algorithmic trading firms. In the 1980s, there were human faces to the successful money managers. Now, it is algorithms managing pension funds, retirement funds, and sovereign wealth funds. Will something similar happen in politics? The appearance of these algorithms in finance went largely unnoticed, and was mostly ridiculed if noticed, when they first appeared quietly almost two decades ago. But now they are used widely and are highly sought after. Will there be a new political party, perhaps, founded by mathematicians and computer scientists, who will approach political problems in a dispassionate manner using algorithms and AI? Will the future Obama, Clinton, Trump, and Cruz be just different types of algorithms?

This might appear to be pure fantasy. But consider this: AI is already making an appearance in the C-suite, helping make executive decisions. Marc Benioff, the CEO of Salesforce, holds weekly meetings with senior executives to talk about their weekly progress and to analyze the quarter's performance. One of the important members present in the weekly meetings is Einstein, an internal Salesforce AI. It is not unusual for Benioff to turn to Einstein for a second opinion about an issue right after a senior executive has spoken. And, interestingly, it is also not unusual for Einstein to single out a specific senior executive for being inaccurate or ambiguous. Now, imagine the same thing happening in politics.

Imagine *The President of the United States*, with an AI assistant, and the AI singling out senior White House staff advisors for her or his deficiencies.

Machine learning and algorithms have, of course, already made an appearance on the political scene. Obama's 2008 campaign was widely known to be driven by data analytics. In 2012, his campaign team used data analytics to make optimal use of a limited budget to target the right voters. More recently, other world leaders, such as Canada's Justin Trudeau[23] and India's Narendra Modi,[24] have been known to use data analytics and tools to help win elections.

The 2016 US presidential election saw even wider usage of data and algorithms. Ted Cruz's team used highly targeted ads on Facebook to win supporters, and used algorithms to label voters as "stoic traditionalists", "temperamental consevatives" or "true believers" to help volunteers fine-tune campaign efforts. Hillary Clinton used help from the data analytics campaign management firm, BlueLabs, founded by senior members of the Obama for America analytics team. Donald Trump's campaign was made highly effective by the data science firm, Cambridge Analytica, the same firm that helped people make up their mind about Brexit.

At least some of the technology used in recent election campaigns seems to have been inspired by recent developments in psychometrics, a data-driven branch of psychology. In the 1980s, psychologists developed the Five Factor Model,[25] also known as the Big Five, according to which any subject's personality traits can be accurately determined based on five factors: openness to experience, conscientiousness, extraversion, agreeableness, and neuroticism. Once a person's standing was known in this five-dimensional space, a lot could be predicted about the person's other characteristics. The problem with this approach was data gathering, It was not easy to collect such detailed data about a person.

Then, around 2010, Michal Kosinski and David Stillwell, fellow doctoral students at Cambridge University, adapted and modified the Big Five model for social media websites. People leave behind an enormous amount of digital footprint in various social media websites. Kosinski and Stillwell published a series of research papers that used machine learning methods on online data from a large number of users, data that was obtained with their consent, and that accurately predicted their race, gender, sexual orientation, religious affiliation, political leaning, and even drug use. The level of accuracy in these predictions was very high, north of 80 percent in many cases. One can only imagine how effective these tools and data sets might have been in the hands of campaign analytics firms.

Isaac Asimov's *Foundation and Robot* sci-fi series revolved around the science of Psychohistory, a fictional algorithmic science that combined history, sociology, and mathematical statistics to make predictions about groups of people. It almost seems like reality is fast catching up with science fiction as we now talk of Big Data Psychometrics and its use in the Brexit referendum and the US presidential election. So, we ask again, is it too early to start worrying about the possibility of AI politicians?

Notes

1 Tom Chatfield is a British author and technology expert. His first book, *FunInc*, was published in 2010. He is a frequent speaker and consultant on technology and media.
2 Enrique Martínez-García, *Globalization: The Elephant in the Room That Is No More*, Federal Reserve Bank of Dallas, Globalization and Monetary Policy Institute 2014 Annual Report. www.dallasfed.org/assets/documents/institute/annual/2014/annua l14b.pdf (Accessed on November 22, 2017)
3 Speech by Richard W. Fisher, President and CEO (2005–2015), Federal Reserve Bank of Dallas, from Globalization and Government Policy, Remarks at the Fifth Annual Federal Reserve Bank of Philadelphia Policy Forum, November 3, 2005 (Harvard University, Cambridge) www.dallasfed.org/news/speeches/fisher/2005/fs051202.aspx (Accessed on November 22, 2017)
4 Brexit is the term used for the decision by the United Kingdom (UK) to withdraw from the European Union. The British government led by David Cameron held a referendum on the issue on June 23, 2016 and the majority voted to leave the European Union. As of the time of writing, the UK remains a full member of the European Union as the terms of withdrawal are being negotiated.
5 The United States presidential election of 2016 was the contest between Republican Donald Trump and Democrat Hillary Clinton. Trump emerged victorious and took office as the 45th President on January 20, 2017.
6 The 2017 French presidential election was held in April and May 2017. As no candidate won a majority in the first round, a run-off was held between the top two candidates, Emmanuel Macron and Marine Le Pen, which Macron won. The presidential election was followed by legislative elections to elect members of the National Assembly in June.
7 "Brexit vote was an unexpected outcome". www.cnbc.com/2016/07/04/why-the-ma jority-of-brexit-polls-were-wrong.html, www.businessinsider.in/Pollsters-now-know-why-they-were-wrong-about-Brexit/articleshow/53363062.cms (Accessed on November 22, 2017)
8 "Trump was trailing Hillary Clinton". www.usnews.com/opinion/op-ed/articles/ 2016-11-10/why-the-polls-got-donald-trumps-2016-win-wrong, http://www.pewre search.org/fact-tank/2016/11/09/why-2016-election-polls-missed-their-mark/ (Accessed on November 22, 2017)
9 Angel Ubide, "Don't blame globalization for the squeezing of the middle class", Centre for European Policy Studies. http://aei.pitt.edu/7361/2/7361.pdf, Pranab Bardhan, "Does globalization help or hurt the world's poor? www.scientificamerican. com/article/does-globalization-help-o-2006-04/ (Accessed on November 22, 2017)
10 Jean-Claude Juncker is a politician from Luxembourg who since 2014 has been President of the European Commission, of the European Union. From 1995 to 2013 he was the Prime Minister of Luxembourg. From 2005 to 2013, Juncker served as the first permanent President of the Eurogroup.
11 "The quest for prosperity – Europe's economy has been underperforming. But whose fault is that?", *The Economist*, March 15, 2007. www.economist.com/node/8808044 (Accessed on November 22, 2017)
12 Gautama Buddha was an ascetic, on whose teachings Buddhism was founded. He is believed to have lived and taught mostly in the eastern part of ancient India sometime between the sixth and fourth centuries BC. Ananda was Buddha's disciple.
13 The French Revolution was a period of immense social and political upheaval in France that began in 1789. The Revolution overthrew the monarchy and established a republic. It was a violent period of turmoil and inspired by liberal and radical ideas. The Revolution profoundly altered the course of European history.

14 The Russian Revolution of 1917 refers to events in Russia during that year which saw two revolutions, the first of which, in February (March, New Style), overthrew the imperial government and the second of which, in October (November), placed the Bolsheviks in power. www.britannica.com/event/Russian-Revolution-of-1917 (Accessed on June 24, 2018)

15 *The Productivity–Pay Gap*, Economic Policy Institute (August 2016). The report discusses *Productivity growth and hourly compensation growth, 1948–2015* in the United States. Generally it is assumed that as the economy expands, everybody should reap the rewards. And for two-and-a-half decades beginning in the late 1940s, this was how the US economy worked. Over this period, the pay (wages and benefits) of typical workers rose in tandem with productivity (how much workers produce per hour). In other words, as the economy became more efficient and expanded, everyday Americans benefited correspondingly through better pay. But in the 1970s, this started to change. The gap between productivity and a typical worker's compensation has increased dramatically since 1973. www.epi.org/productivity-pay-gap/ (Accessed on November 22, 2017)

16 A chart first published in a 2012 World Bank working paper by economist Branko Milanovic details which segments of the global population saw a rise in real incomes from 1988 to 2008. The benefit to consumers in advanced economies took the form of downward price pressures on goods produced at cheaper locations. But, the middle classes in developed nations failed to benefit much. The biggest losers (other than the very poorest 5 percent) of globalization were those between the 75th and 90th percentiles of the global income whose real income gains were essentially nil, according to Milanovic. http://documents.worldbank.org/curated/en/959251468176687085/pdf/wps6259.pdf, www.bloomberg.com/news/articles/2016-06-27/get-ready-to-see-this-globalization-elephant-chart-over-and-over-again (Accessed on November 22, 2017)

17 "Artificial intelligence: The impact on jobs; Automation and anxiety - Will smarter machines cause mass unemployment?" *The Economist* Special report. www.economist.com/news/special-report/21700758-will-smarter-machines-cause-mass-unemployment-automation-and-anxiety (Accessed on November 22, 2017)

18 David Autor, David Dorn, Gordon Hanson and Kaveh Majlesi, "Importing political polarization? The electoral consequences of rising trade exposure", Working Paper 22637. Cambridge, MA: National Bureau of Economic Resarch, September 2016. www.nber.org/papers/w22637, https://seii.mit.edu/wp-content/uploads/2016/04/SEII-Research-2016-Autor-Dorn-Hanson-Majlesi.pdf, www.minneapolisfed.org/publications/the-region/interview-with-david-autor (Accessed on November 22, 2017)

19 Daron Acemoglu, David Autor, David Dorn, Gordon H. Hanson and Brendan Price. 2016. "Import competition and the great US employment sag of the 2000s". *Journal of Labor Economics*, 34(S1): S141–S198. https://economics.mit.edu/files/9811 (Accessed on November 22, 2017)

20 Paul Tudor Jones II is an American investor, hedge fund manager, and philanthropist. In 1980, he founded his hedge fund, Tudor Investment Corporation, an asset management firm in Greenwich, Connecticut. As of February 2017, *Forbes Magazine* estimated his net worth to be US$4.7 billion, making him the 120th richest person on the Forbes 400. He is known for his large-scale philanthropy and focuses on poverty reduction.

21 George Soros is an American investor, philanthropist and author. Soros is considered to be one of the most successful investors in the world. As of May 2017, Soros has a net worth of $25.2 billion, making him one of the 30 richest people in the world. He is known as "The Man Who Broke the Bank of England" because of his short sale of US $10 billion worth of Pound sterling, making him a profit of $1 billion during the 1992 Black Wednesday UK currency crisis.

22 Bruce Stanley Kovner is an American investor, hedge fund manager and philanthropist. He is Chairman of CAM Capital, which he established in January 2012 to manage his investment, trading and business activities. From 1983 through 2011,

Kovner was Founder and Chairman of Caxton Associates, LP, a diversified trading company.

23 Justin Pierre James Trudeau is a Canadian politician. He is the 23rd and current Prime Minister of Canada and leader of the Liberal Party. The second youngest prime minister after Joe Clark, he is also, as the eldest son of former Prime Minister Pierre Trudeau, the first to be related to a previous holder of the post.

24 Narendra Modi is an Indian politician who is the 14th and current Prime Minister of India. He has been in office since 26 May 2014. Modi is a prominent leader of the Bharatiya Janata Party (BJP) and was also the Chief Minister of Gujarat from 2001 to 2014, and is the Member of Parliament from Varanasi in Uttar Pradesh. Modi led the BJP in the 2014 general election, which gave the party a majority in the Lok Sabha (the Lower House of Indian Parliament), the first time a single party had achieved this since 1984.

25 The Five Factor Model is a model based on common language descriptors of personality. This theory suggests five broad dimensions used by psychologists to describe the human personality and psyche. The five factors have been defined as openness to experience, conscientiousness, extraversion, agreeableness and neuroticism, often listed under the acronyms OCEAN or CANOE.

EPILOGUE

> The saddest aspect of life right now is that science gathers knowledge faster than society gathers wisdom.
>
> —*Isaac Asimov*[1]

George Orwell, *Animal Farm*, and AI

George Orwell's famous novel *Animal Farm*[2] has an interesting beginning and an even more interesting end. At the beginning of the novel, the animals revolt against the irresponsible and cruel owner of a farm and take control over it. The pigs were leaders of the revolution and they take up key roles. But as the years pass, the farm becomes a personal fiefdom of a coterie. However, we are not talking about the merits and demerits of one system versus another here.

It is the end of the novel that is very interesting. Slowly, the pigs start to resemble humans in all possible ways, e.g. they walk upright and wear clothes. There is also a vast difference between what they practice and what the leaders at the farm preach. Towards the end of the novel, the in-charge of the farm holds a dinner for the pigs and local farmers to celebrate a new alliance. The name of the farm is changed to the original once again and essentially, this was confirmation that some animals become more equal than others. As the other animals on the farm look from pigs to humans, they realize that they can no longer distinguish between the two.

Is this what the end will look like? Will AI machines and humans become indistinguishable? Will humans finally become subservient to AI machines? Or, will AI turn out to be a leap of unprecedented progress for humankind and far superior to any other technology we have developed so far? Will AI be the harbinger of a new and better-than-ever era for humankind? Or, have we just overhyped the impact of AI in terms of its impact on jobs and everything else and AI will not be able to live up to the promise, and we will be heading soon towards a fresh AI winter?

We may not have all answers but AI is different and disruptive

Today, we don't have all the answers. But, there is no doubt that AI is a disruptive force and it certainly has the potential to become a powerful game changer. We understand that "disruptive" or "disruption" are much-abused words in today's world, especially in the domain of technology. However, AI technology has already proved that there are times when it is disruptive in a real sense. Ask the brick and mortar stores about Amazon or the licensed taxi drivers about Uber.

The supporters of AI claim that these concerns are unfounded as doomsday scenarios have been proven wrong umpteen numbers of times throughout history. The tractor was supposed to render thousands of workers jobless, the automobile was a threat to drivers of horse carriages and similarly, thousands of people were supposed to lose their jobs when computers came on the horizon. However, millions of new jobs were created and humankind witnessed the advent of several new industries whenever these changes happened.

However, this time may really be different. We have discussed in Chapter 3, *Emerging danger of AI-induced mass unemployment* that AI is making a significant impact on jobs and through this on society. We simply cannot view and treat AI as just another technology or scientific advancement that humans have achieved. We need to have a comprehensive policy response to deal with the changes that will be brought by AI. We need think tanks and policy makers to really analyze, understand and talk about the issues. We need the corporate world to think about the ethics of AI. We need the government to think about the possible impact on the livelihood of people because of AI and the actions that need to be taken to mitigate the impact.

The change is already visible. A number of smart home devices now use some basic form of AI. By setting defaults and preferences, the indoor lighting might adjust depending on where you and what you are doing. Siri and Google Voice can understand instructions and follow through. Recommender systems in Netflix or Pandora can learn your preferences from your previous choices and make good recommendations for movies or music. News writing bots can now write simple reporting pieces at news agencies such as AP, Yahoo and Fox.[3] You might be getting help from an online customer support member at your favorite website without even realizing that the helper is really a AI bot. Fraud detection, purchase prediction, smart cars, the list goes on. Change is ubiquitous though we may not always notice it when it is continuous and slow.

AI and parallels with nuclear technology

There is an interesting and relevant example from the past. The AI researchers and scientists will probably need to take a leaf out of how nuclear technology was developed. One of the most challenging parts in nuclear energy is how to always keep nuclear fission as a controlled reaction. That's the thin line between whether nuclear technology will be safe or whether it will be a disaster because initiating a nuclear reaction is much easier than keeping it controlled.

When the nuclear reaction is controlled, it is a source of energy and has several other beneficial uses such as in medicine. Over the last 70 years, nuclear energy has helped humanity to a great extent. However, when the nuclear reaction becomes uncontrolled, we get massive disasters in the form of Hiroshima, Nagasaki,[4] Chernobyl[5] and Fukushima.[6] The difference between Hiroshima and Nagasaki vs. Chernobyl and Fukushima is that while humans may have taken a terrible step in 1945, they were still in control and the action was deliberate. But in Chernobyl and Fukushima, humans were not in control.

And fortunately, humankind did get a second chance with nuclear bombs and there has been no repeat in more than 70 years. In the case of Chernobyl and Fukushima, the damage was considerably less than what could have been the worst-case scenario and this was only because we were lucky. There is no certainty that we will be as fortunate with AI. If AI becomes uncontrollable, that would be the end of the story, literally.

Would you want Mark Zuckerberg to be the next US President?

The question is incomplete. We are not asking if you want the founder of Facebook to be the next US President. We don't know if he is even interested; we also don't know how he will compare with the presidents in the past or the candidates in previous presidential elections. We may know a bit about his views on some of the important issues but, mostly, we don't know much about what he thinks on all the important issues that matter for a US President.

Ideally, anyone who is eligible to contest the elections and is keen should stand a chance to get elected. The person should get a fair chance to put forward his or her views and then should be evaluated as to whether he or she is fit to be a president. It is not necessary that the presidential candidates come from a political background. Just because Zuckerberg is rich or leads Facebook does not make him ineligible. So, the question here is not if he is interested or if he will contest or if he will get elected.

We are in fact asking if Mark Zuckerberg of Facebook will be the next US President because he controls Facebook. If he decides to contest and if he does not find it morally or ethically wrong to use Facebook to influence public opinion, will there be a mechanism available to stop him? If Facebook decides to profile the voters on the basis of what they do on Facebook and then voters are selectively fed the information that is going to tilt the scales in favor of Zuckerberg, what are the consequences then for democracy?

We are not even talking about fake news or post truth, we are only saying that it is a fact that Facebook knows a lot about its users in terms of what they like, what they don't, what they are interested in, their views on political and social issues. If that information on voters can be used to influence the voters' decision in someone's favor or against someone, will it be equivalent to compromising democracy? Perfectly genuine information and nothing but the truth can also be presented in such a way that it may lead to selective listening.

There are examples in the history of US presidential elections when industrialists and people heading giant corporations had a desire to contest. They had huge resources at their disposal and could use them to further their political interests. How is Mark Zuckerberg's case different? Here the difference lies in the fact that while rich candidates can spend a lot of money, this does not guarantee that they will be able to change the thought process of voters.

On the other hand, Facebook or Google have the ability to tinker with the thought process. Arguably, tinkering with the thought process is an unfair and unnatural advantage while the ability to spend money that is earned and owned by the candidate is a natural advantage. This may not be a pleasant thought for people who believe that democracy should provide an equal opportunity for ordinary people to run for the highest political office in the United States of America.

Impact of AI on human emotions and human relationships

There are people who claim, and of course they have some solid arguments, that technology and scientific progress is negatively impacting the emotional responses of humans, while others take an opposite view and say that because of scientific progress, we are kinder and much more understanding now. There are also arguments and counter-arguments over how technology has altered human relationships and people have become more and more individualistic and less tolerant to divergent views. Since AI is a different kind of technology to all others we have seen previously, the concerns are much greater that AI will have a severe impact on human emotions and human relationships.

Despite what is being portrayed in popular culture and how the "robots who can think and who have emotions" are shown in movies, which of course is a gross exaggeration at times, it is not unlikely that there would be a significant impact on human relations and emotions because of machines that can think like humans. While it cannot be conclusively proven whether technology has made us less human or more human, one thing is certain. With scientific innovations, humans have evolved in their behavior and emotional responses. As they saw faster and faster rises in material well-being, they changed in response to this transformation.

TABLE E.1 Maslow's hierarchy of human needs

Hierarchy	*Need classification*	*Examples for specific needs inclusion*
1	Survival needs	Breathing, water, food, sleep, clothing, shelter
2	Safety and security needs	Personal security, financial security, well-being
3	Love and belonging needs	Family, friends and need to feel loved
4	Self-esteem needs	Respect and recognition
5	Self-actualization needs	Realization of potential

Source: www.simplypsychology.org/maslow.html (Accessed on 12 April 2018), www.psychologytoday.com/us/blog/hide-and-seek/201205/our-hierarchy-needs (Accessed on 12 April 2018)

The instinct to survive may not have changed but as resources increased and health standards improved, the struggle involved in survival has reduced considerably. Over last 60–70 years, technology has grown by leaps and bounds and in entirely new areas of Information Technology and Communication. But why did it happen that on the one side, the world has been getting more peaceful, healthier and prosperous, and on the other, people have been becoming unhappy, dissatisfied and distressed? Perhaps Maslow's theory on a hierarchy of human needs provides an interesting point of view.[7]

Abraham Maslow's theory of human needs

As Maslow highlighted, with every lower-level need fulfillment, human beings move to the next level of need. As prosperity increases, society starts taking the fulfillment of lower-level needs for granted. As living comfort goes up, so does the need for belonging and afterwards, the need for self-esteem and self-actualization, respectively. For most of human history, humans were so engrossed with meeting the needs for survival and safety that only very few people may have thought of the need for self-esteem and self-actualization. When people have more free time after their survival and safety needs are met, they focus more on what is missing in their lives in terms of relationships, self-esteem and an unfulfilled realization of potential.

The shift from REAL to VIRTUAL and impact on psychological health

In an interesting paper titled *Prevalence of perceived stress, symptoms of depression and sleep disturbances in relation to information and communication technology (ICT) use among young adults – an explorative prospective study*, the researchers looked at whether a high quantity of ICT[8] use is a risk factor for developing psychological symptoms among young users. A group of college students was assessed on different types of ICT use and perceived stress, symptoms of depression and sleep disturbances.

This paper also suggests that for women, a high combined use of computer and mobile phone at baseline was associated with increased risk of reporting prolonged stress and symptoms of depression at follow-up, and the number of short message service (SMS) messages per day was associated with prolonged stress. Also, online chatting was associated with prolonged stress, and emailing and online chatting were associated with symptoms of depression, while internet surfing increased the risk of developing sleep disturbances. The findings clearly suggest that ICT may have an impact on psychological health.

There is more evidence. Another research paper,[9] *Association between media use in adolescence and depression in young adulthood—A longitudinal study*, studies the association between media exposure and depression. The objective of the study was to assess the longitudinal association between media exposure in adolescence and depression in young adulthood in a nationally representative sample. It was found in the study that people with more media exposure had significantly greater odds of developing depression. The conclusion was that television exposure and total media exposure in adolescence are associated with increased odds of depressive symptoms in young adulthood, especially in young men.

There are also more localized studies. In another research study,[10] *A prospective study of screen time in adolescence and depression symptoms in young adulthood*, the researchers examines the association between screen time in adolescence and depressive symptoms in young adulthood among Danish adolescents. It was found that limiting screen time, particularly television viewing, during adolescence may be important for preventing depression in young adulthood. This was an important finding because only when you are able to establish a definite link between habits or behavior and mental health issues, will you be able to take preemptive action.

In another research study, *Association between screen time and depression among US adults*,[11] surveys conducted in general populations have found that the prevalence of depression is about 9 percent in the United States and the objective of this study was to assess the relationship between television watching/computer use and depression. In this study, depression was found to be significantly higher among female and the results showed that a moderate or severe depression level was associated with a higher amount of time spent watching the TV and using the computer. The conclusion of the study was that TV watching and computer use can predict the depression level among adults.

Just like the impact of television, smartphones, internet and other technologies such as virtual reality, there is a very high likelihood that AI, being the technology it is, will lead to a deep psychological impact on the newer generations. Whether AI will be good or bad for us and how good or how bad it will be, only time will be able to tell. But this is an important dimension of AI that will have significant social implications. It would help if we are sensitive about these issues and try to see them in totality without being in a hurry to declare AI as simply good or bad.

Public debate and popular opinion have an irreplaceable role

The recent controversy about the personal data leak of millions of Facebook users and the suspicious role of Cambridge Analytica has brought the debate back to the fore on how online activity is being used by technological firms to not only sell but also to influence public opinion. The use of personal data for devising targeted campaigns and to mold public opinion is going to be much more harmful than personalized selling. When you pay more for something you are buying, or even worse, buy something because you are being nudged, the damage is limited to an individual. But when you are influencing the voting choices and trying to shape public opinion, the damage is far greater because it collectively affects the entire society.

One bad choice today could harm generations to come. This is not to claim that humankind has always been making the right choices about the future but at least our decision-making was not being impacted because there was somebody invisible working behind the scenes to do that without our knowledge. Without us being aware that we are being fed information selectively, we will never have the complete picture and information asymmetry has always been a major cause of poor decisions. The decisions may be poor even otherwise, but when a new variable is sneaking in and we are not aware, the entire process gets compromised. The worst part is that people will not even know that they have been taken for a ride. Is there anything worse than being robbed? Yes, there is and it is called being robbed without you even knowing.

The other big issue is about the lack of control. As the Facebook-Cambridge Analytica fiasco has shown, even when you believe the most benign explanation that has been offered and have complete and absolute trust in Mark Zuckerberg's testimony at US Congress, the idea that your personal data is being used against you so that some crooked politicians can accumulate more power is extremely disturbing. When the level of debate and quality of public discourse is already stooping lower and lower, it makes sense to remain even more cautious that at least the levers of decision-making are not surrendered to companies and people working for nothing else but their own selfish ends.

Whether the new developments in AI will create more inequalities between societies and countries, leading to more discord, or whether they will lead to more prosperity and peace, depends on what we do today. It is likely that the stakeholders will feel obliged to behave responsibly if the discussion on AI becomes prominent in the mainstream and the role of AI takes center stage in frontline media. The issue has been sidelined for some time, but we cannot postpone any longer fully addressing the questions. The public has to get involved and the opinion makers have a big role to play in it. In our effort through this book, if we are helpful, even to a small degree, in doing the same, we will be happy that we have done our job well.

Notes

1 This quote is from Isaac Asimov and Jason A. Shulman's *Isaac Asimov's Book of Science and Nature Questions* (New York: Weidenfeld & Nicolson, 1988), p. 281.—Themis-Athena (talk) 10:02, November 16, 2012 (UTC).

2 George Orwell was born Eric Arthur Blair in 1903 near India's border with Nepal. Orwell's father worked as an agent of the Indian Civil Service during British rule. George Orwell was an English novelist, essayist, journalist, and critic. Orwell's novel *Animal Farm* was first published in England in 1945. The story is based on events around the Russian Revolution of 1917 and, later on, from the Stalin era in the Soviet Union. *Time* magazine chose the book as one of the 100 best English-language novels (1923 to 2005) and it also featured on the "Modern Library List of Best 20th-Century Novels". The basic premise of *Animal Farm* is the theoretical concept of exploitation of workers in individual ownership under capitalism vs. benefit for all in collective ownership under socialism.

3 Klint Finley, "This news-writing bot is now free for everyone", *Wired Business*, Oct. 20, 2015. www.wired.com/2015/10/this-news-writing-bot-is-now-free-for-everyone/ (Accessed on November 20, 2017)

4 During the final stage of World War II, the United States of America dropped nuclear weapons on the Japanese cities of Hiroshima and Nagasaki on August 6 and 9, 1945, respectively. The United States had dropped the bombs with the consent of the United Kingdom. The two bombings killed at least 129,000 people and remain the only use of nuclear weapons in history.

5 The Chernobyl disaster, also referred to as the Chernobyl accident, was a catastrophic nuclear accident. It occurred on April 26, 1986 in the light water graphite moderated reactor at the Chernobyl Nuclear Power Plant near Pripyat, in what was then part of the Ukrainian Soviet Socialist Republic of the Soviet Union (USSR). A combination of inherent reactor design flaws and the reactor operators arranging the core in a manner contrary to the checklist resulted in an uncontrolled reaction. Water flashed into steam, generating a destructive steam explosion and a subsequent open-air graphite fire. This fire produced considerable updrafts for about nine days.

6 The Fukushima Daiichi nuclear disaster was an energy accident at the Fukushima Daiichi Nuclear Power Plant in Fukushima, initiated primarily by the tsunami following an earthquake in Japan on March 11, 2011. Immediately after the earthquake, the active reactors automatically shut down sustained fission reactions. However, the tsunami disabled the emergency generators that would have provided power to control and operate the pumps necessary to cool the reactors. The insufficient cooling led to nuclear meltdowns, hydrogen-air explosions, and release of radioactive material.

7 Maslow's hierarchy of needs is a theory in psychology proposed by Abraham Maslow in his 1943 paper "A Theory of Human Motivation". Maslow subsequently extended the idea to include his observations of humans' innate curiosity. His theories parallel many other theories of human developmental psychology, some of which focus on describing the stages of growth in humans. Maslow used the terms "physiological", "safety", "belonging and love", "esteem", "self-actualization", and "self-transcendence" to describe the pattern that human motivations generally move through. The goal of Maslow's theory is to attain the sixth level or stage: self-transcendent needs. Maslow's theory on need hierarchy remains a popular framework in sociology research, management training and secondary and higher psychology.

8 Sara Thomée, Mats Eklöf, Ewa Gustafsson, Ralph Nilsson, Mats Hagberg, 2007. "Prevalence of perceived stress, symptoms of depression and sleep disturbances in relation to information and communication technology (ICT) use among young adults – an explorative prospective study". *Computers in Human Behavior*, 23(3): 1300–21. https://doi.org/10.1016/j.chb.2004.12.007 (Sara Thomée is based at Occupational and Environmental Medicine, Sahlgrenska Academy and University Hospital, Göteborg, Sweden). www.sciencedirect.com/science/article/pii/S0747563204002250 (Accessed on January 25, 2018), www.sciencedirect.com/science/article/pii/S0747563204002250?via%3Dihub (Accessed on January 25, 2018)

9 Brian A. Primack, Brandi Swanier, Anna M. Georgiopoulos, Stephanie R. Land, & Michael J. Fine. 2009. "Association between media use in adolescence and depression in young adulthood: A longitudinal study". *Archives of General Psychiatry*, 66(2): 181–88. www.ncbi.nlm.nih.gov/pmc/articles/PMC3004674/?iframe=true&width=100%25&height=100%25 (Accessed on January 25, 2018)

10 A. Grøntved, J. Singhammer, K. Froberg, N. C. Møller, A. Pan, K. A. Pfeiffer, P. L. Kristensen. 2015. "A prospective study of screen time in adolescence and depression symptoms in young adulthood". *Preventative Medicine*, 81: 108–13.www.ncbi.nlm.nih.gov/pubmed/26303369 (Accessed on January 25, 2018)

11 K. C. Madhav, S. P. Sherchand, & S. Sherchan/ 2017. "Association between screen time and depression among US adults". *Preventative Medicine Reports*, 8: 67–71. doi: 10.1016/j.pmedr.2017.08.005. eCollection Dec. 2017. www.ncbi.nlm.nih.gov/pubmed/28879072 (Accessed on January 25, 2018)

INDEX